Service Robotics within the Digital Home

AF213196

International Series on
INTELLIGENT SYSTEMS, CONTROL AND AUTOMATION:
SCIENCE AND ENGINEERING

VOLUME 53

Editor

Professor S. G. Tzafestas, National Technical University of Athens, Greece

Editorial Advisory Board
Professor P. Antsaklis, University of Notre Dame, Notre Dame, IN, USA
Professor P. Borne, Ecole Centrale de Lille, Lille, France
Professor D.G. Caldwell, University of Salford, Salford, UK
Professor C.S. Chen, University of Akron, Akron, Ohio, USA
Professor T. Fukuda, Nagoya University, Nagoya, Japan
Professor S. Monaco, University La Sapienza, Rome, Italy
Professor G. Schmidt, Technical University of Munich, Munich, Germany
Professor S.G. Tzafestas, National Technical University of Athens, Athens, Greece
Professor F. Harashima, University of Tokyo, Tokyo, Japan
Professor N.K. Sinha, McMaster University, Hamilton, Ontario, Canada
Professor D. Tabak, George Mason University, Fairfax, Virginia, USA
Professor K. Valavanis, University of Denver, Denver, Colorado, USA

For other titles published in this series, go to
http://www.springer.com/series/6259

Ignacio González Alonso • Mercedes Fernández
José M. Maestre
María del Pilar Almudena García Fuente

Service Robotics within the Digital Home

Applications and Future Prospects

Springer

Ignacio González Alonso
Infobotica Research Group
Computer Science
Oviedo University
Calle González Quirós s/n
33600 Mieres, Asturias
Spain
gonzalezaloignacio@uniovi.es

Mercedes Fernández
Infobotica Research Group
Computer Science
Oviedo University
Calle González Quirós s/n
33600 Mieres, Asturias
Spain
fernandezmercedes@uniovi.es

José M. Maestre
Department of Systems and Automation
Engineering
University of Seville
Descubrimientos s/n
41092 Sevilla
Spain
pepemaestre@cartuja.us.es

María del Pilar Almudena García Fuente
Infobotica Research Group
Computer Science
Oviedo University
Calle González Quirós s/n
33600 Mieres, Asturias
Spain
agarciaf@uniovi.es

ISBN 978-94-024-0547-7 ISBN 978-94-007-1491-5 (eBook)
DOI 10.1007/978-94-007-1491-5
Springer Dordrecht Heidelberg London New York

© Springer Science+Business Media B.V. 2011
Softcover re-print of the Hardcover 1st edition 2011
No part of this work may be reproduced, stored in a retrieval system, or transmitted in any form or by any means, electronic, mechanical, photocopying, microfilming, recording or otherwise, without written permission from the Publisher, with the exception of any material supplied specifically for the purpose of being entered and executed on a computer system, for exclusive use by the purchaser of the work.

Cover design: eStudio Calamar S.L.

Printed on acid-free paper

Springer is part of Springer Science+Business Media (www.springer.com)

Preface

Over the past few decades there has been an exponential growth in service robots and smart home technologies, which has led to the development of exciting new products in our daily lives. Service robots can be used to provide domestic aid for the elderly and disabled, serving various functions ranging from cleaning to entertainment. Service robots are divided by functions, such as personal robots, field robots, security robots, healthcare robots, medical robots, rehabilitation robots and entertainment robots. A smart home appears "intelligent" because its embedded computers can monitor so many aspects of the daily lives of householders. For example, the refrigerator may be able to monitor its contents, suggest healthy alternatives and order groceries. Also, the smart home system may be able to clean the house and water the plants.

However, the operation of all these devices, systems or robots does not really require them to "think" as they are simply programmed to perform a series of repetitive tasks. If anything interferes with the pre-programmed task, they would malfunction since none of them is able to sense the interference and think out a solution. As service robots are in greater proximity to humans, the technology involves more safety concerns over the human–machine interaction. Therefore, it remains a great challenge today for us to build smart homes and intelligent robots that can "think" like we do.

To achieve such a goal, scientists and engineers have been trying hard to capture the essence of human intelligence in our homes and robots to make them intelligent to function well in the real world. This is a challenging and ambitious task since the robot or home intelligence must cope with various noises, uncertainty and dynamic changes in the real world. Like human beings, smart homes and intelligent robots should be able to sense their environments, reason and make decisions and respond to tasks and unexpected events quickly.

In general, intelligent robots and smart homes are broad, interdisciplinary subjects that involve many different technologies such as sensor integration, data fusion, wireless sensor networks, map building, embedded computing, navigation, planning and artificial intelligence.

For each individual smart home or robot, the "thinking" process takes place at many different levels. At its lowest level, "thinking" needs to be fast to respond quickly to unexpected events. At higher levels, "thinking" enables the homes and robots to handle a dynamic and uncertain world. By contrast, "thinking" should exhibit an adaptation and learning capability at its highest level. Moreover, a close interaction among smart homes and robots should be made to achieve the common goal cooperatively.

This book aims to address some fundamental issues related to the integration of robots into smart homes. To build a home or a robot that is completely autonomous and truly intelligent was only a dream yesterday, as reflected by science fiction books and films such as *I Robot* and *Dr. Who*. It remains a dream today since most of the autonomous robots and smart homes currently being built do not function well in the real world. This is mainly attributed to our incomplete understanding of the process of perception, recognition and reasoning in humans, the limitations of available scientific methods and the incompatibility of today's computer technology. However, our dream will come true as our research efforts continue in the twenty-first century. Although it is difficult and often unwise to predict the future, we believe that we are gradually progressing in that direction as faster computers and new sensor technology become available. The future is bright. Let us face the challenge and work together towards the integration of smart homes and intelligent robots step by step.

University of Essex, UK Professor Huosheng Hu

Acknowledgments

All rights reserved. No part of the contents of this book may be reproduced or transmitted in any form or by any means without the written permission of the publisher.

Ministerio de Industria, Turismo y Comercio
http://www.mityc.es

This work was possible thanks to the financial support of the Spanish Department of Science and Technology. We must acknowledge the continuous support given by the University of Oviedo in providing all the resources that made possible our research and, in particular, by their management of the DHCompliant project grant: MITC-09-TSI-020100-2009-359.

Some companies have participated like partners, so we express our gratitude to them.

Domotica DaVinci
www.domoticadavinci.com

Centro Tecnológico Cartif
www.cartif.com

Ingenium
www.ingeniumsl.com

MoviRobotics
www.movirobotics.com

Sevilla University
www.us.es

Various organizations have assisted with the production of this book by providing images for use in the figures. The authors wish to acknowledge their support. These organizations include (in alphabetical order):

América Kyodo
ANYBOTS
Applied Research Associates, Inc.
CoroWare Inc.
Cyberbotics
Evolution Robotics
Inderscience Enterprises Limited
Infobótica Research Group
Int. Spont. Net (Michael Jeronimo)
KA Home Robotics
Laurent Ricatti – AnyKode Marilou
NASA
National Institute of Standards and Technology
Rey Juan Carlos University
Robocup
Spykee
The Joef Stefan Institute
University of Seville
University of Western Australia
UPnP Forum
Willowgarage
YujinRobot

Acknowledgments by Author

Ignacio González Alonso
To my sweet Veronica,
To Rocio Fernández Cifuentes,
To my family and friends, and all the people from the Infobotica Research Group.

Mercedes R. Fernández Alcalá
To my family,
To (Y.C.S & T.L.M) and all my friends
To J.A.L.B.

José M. Maestre
In first place, I would like to thank all the people who have worked with me in this field at the University of Seville. More specifically, I would like to recognize the support of A. Álvarez, J. R. de la Pinta, C. Martín and E. F. Camacho. Likewise, I would like to thank my family for their continuous love and support. In particular, I would like to dedicate my contribution to my brother, with the hope that he will help my interoperability dream come true when he becomes a computer science engineer.

María del Pilar Almudena García Fuente
To my mother, no more than one.

Contents

About the Authors

Dr. Ignacio González Alonso (M'09) was born in Oviedo, Spain in 1978. He became a Member (M) of IEEE in 2009. He graduated as a computer science engineer (University of Oviedo, Spain, 2002); he also studied for his Master's in Informatics engineering (University of Oviedo, Spain, 2006); in 2009 he was awarded a Ph.D. in Computer Science with a thesis on generative programming.

He was the co-founder of Criptonet (2001–2002) and Negocios y Robótica. Since 2005, he has held an Assistant Professorship at the University of Oviedo, Spain. Moreover, he is the first author of the paper that received a best paper award at the ICONS2010 conference. He is also the author of "Robots in the smart home: a project towards interoperability." He is interested in the fields of interoperability of heterogeneous systems, model-driven systems engineering, human–robot interfaces, environmental technologies and energy management.

Dr. Gonzalez Alonso is associated with SAE international, INCOSE, EUROP/EURON, OMG, EPoSS, AER-ATP, ASIMELEC and HISPAROB technological platforms.

Dra. Mercedes R. Fernández Alcalá was born in Veracruz, México and is a researcher at the Infobotica Research Group at the University of Oviedo. She obtained her Ph.D. degree in Computer Science in October 2009. She also has a Telecommunication Specialist Degree and a Master's in Corporative Networks and System Integration from Politechnic Valencia University. She was the author of the conference contribution entitled: "Interoperability Standard used by Service Robots" that received a best paper award at the ICONS2010. She is interested in the fields of human-robot interfaces and e-learning technologies among other technologies.

Dr. José M. Maestre is a telecommunications engineer from the University of Seville, where he works as a postdoc researcher. In addition, he received a Master's degree in Smart Home and Building Automation Technologies from the Universidad Politécnica de Madrid in 2006 and a Master's degree in Telecommunication Economics from the Universidad Nacional de Educación a Distancia in 2010. He obtained his PhD in 2010. His research activity is focused on the control of distributed systems and interoperability in smart homes. He has authored and co-authored more than 20 papers in journals and conferences regarding these topics. He is also one of the founders of the firm Idener, a spin-off of the University of Seville.

Dra. María del Pilar Almudena García Fuente graduated as a mining engineer (University of Oviedo, Spain 1987); she earned her Ph.D. in 1996. She holds a Professorship at the University of Oviedo, (Spain, 1989). She has written three books that were published by the publications services of University of Oviedo titled "Introduction to structured programming and Object-Oriented Pascal," "Fortran programming language" and "Introduction to Computers". She is interested in the fields of interoperability of heterogeneous systems, model-driven systems engineering, human–robot interfaces, environmental technologies, energy management, home and building control, agro-forestry and logistics, industrial and manufacturing supportive technologies.

Abbreviations

CARMEN	Carnegie Mellon Navigation Toolkit
CLARAty	Coupled Layer Architecture for Robotic Autonomy
CORBA	Common Object Request Broker Architecture
COU	Control Operator Unit
DHCP	Dynamic Host Configuration Protocol
GENA	General Event Notification Architecture
HTML	Hypertext Markup Language
HTTP	Hypertext Transfer Protocol
IDL	Interface Definition Language
IFR	International Federation of Robotics
MARIE	Mobile and Autonomous Robotics Integration Environment
MDARS	Mobile Detection Assessment Response System
MFC	Microsoft Foundation Class Library
OMG	Object Management Group
OpenGL	Open Graphics Library
OPNET	Optimized Network Engineering Tool
OROCOS	Open RObot COntrol Software
OSGi	Open Services Gateway Initiative
RMI	Java Remote Method Invocation
SOAP	Simple Object Access Protocol
SSDP	Simple Service Discovery Protocol
STL	Standard Template Library
TCP/IP	Transmission Control Protocol/Internet Protocol
UDDI	Universal Description, Discovery and Integration
UDP/IP	User Datagram Protocol/Internet Protocol
VSE	Microsoft Visual Simulation Environment
WS	Web Services
WSDL	Web Services Definition Language
XML	Extensible Markup Language

Introduction

"Any sufficiently advanced technology is indistinguishable from magic."

Clarke, Arthur C.

This book aims to introduce the reader to the exciting world of interoperability between service robots and the digital home. There are enormous differences inside this field, so having the different topics of the book gathered together will help any researcher or developer. The authors' purpose is to help manage the complexity of the development of such systems.

Home and building control, interoperability and service robotics could each require an entire book for themselves, but this book will focus on adding some light to the grey areas between them. Moreover, it will help with the practical matters that the DH Compliant Consortium has already identified in its day-to-day work.

Our first chapter is devoted to interoperability and its standards, followed by the second chapter that describes different development technologies. The third chapter is a compendium of the different service robots that can interoperate in the digital home context or have a strong or potential link with it. Chapter 4 covers home and building control technologies. Finally, Chapter 5 contains an analysis of the field, trends and a short forecast. As any forecast, it has a limited value, but it is included to assist the reader in imagining his or her own forecasts or joining the discussion of the authors' views.

December, 2010 Ignacio González Alonso

Chapter 1
Interoperability Systems

María del Pilar Almudena García Fuente, Javier Ramírez de la Pinta, and Adrián López García

Abstract Since the most important objective of information systems is focused on the development, use and administration of the technology that serves a group of companies, the intercommunication among these information systems must be able to satisfy a large number of needs. This is the main reason for standard information systems that are being used currently. To achieve functionality and interoperability among these systems, markup standards, consulting services and some web services are required. The variety of information systems companies represents the main problem for interoperability among systems, because minimum requirements are not usually established. However, we need to realize that interoperability among systems surpasses simple breakdowns in exchange of information. Our systems also have to, simultaneously, simplify the use of common platforms that can deal with different languages. There are also other important aspects, such as the interaction and implementation of a number of tasks at the same time.

1.1 Introduction

When looking through history from distributed systems to interoperability, it is important to recognize a need for cooperation and expansion of networks that are already in use. The beginning is the concept of *middleware*. Middleware is connectivity

M. del P.A.G. Fuente (✉)
University of Oviedo, Oviedo, Spain
e-mail: agarciaf@uniovi.es

J.R. de la Pinta
Department of Systems and Automation Engineering, University of Seville, Seville, Spain
e-mail: jrdelapinta@cartuja.us.es

A. L. García
Ingenium S.L., Barcelona, Spain
e-mail: adrian@ingeniumsl.com

I.G. Alonso et al., *Service Robotics within the Digital Home*, Intelligent Systems, Control and Automation: Science and Engineering 53, DOI 10.1007/978-94-007-1491-5_1,
© Springer Science+Business Media B.V. 2011

software that offers a group of services that make the running of distributed applications over heterogeneous platforms possible.

The idea of *middleware*, as an abstract layer of software, is to encapsulate all the available resources on a network, which can comprise all kinds of devices (from embedded processors to super processors, laptops, PDAs and mobile phones) and interconnect them in a transparent way. In other words, give an API (Application Programming Interface) to the programmers for the use of distributed applications.

There are some works related to the design and implementation of middleware generic distributed systems. For example, Blair et al. (1998), which represents an approximation to the design of a configurable system, based on the concept of *reflection*. The usefulness of this component's engineering is also important when giving a system the ability to configure and reconfigure.

These concepts are also commented by Coulson et al. (2002). These authors also talk about the link to the application layer by using this component's technology. They suggest the development of a model (OpenORB, based on the model CORBA) independent of the platform and the language of programming. They also define meta-structures and meta-data to give intelligence to the protocol so it can apply reason to its own interpretations, so the system's (re)configuration will be easier.

In this section, the protocols or existing systems used to communicate among heterogeneous platforms will be described, focusing on those based on the concept of ***service***.

1.2 UPnP

1.2.1 Introduction

UPnP (Universal Plug and Play) is a group of protocols or a much-extended architecture suggested by Microsoft (Olleros 2007) and promulgated by the UPnP Forum, which ensures that some network devices can autoconfigure. The aims of UPnP are making sure that the devices can connect perfectly and simplifying the implementation of networks at home (exchange of data, communications and entertainment) and in corporate environments. It is an open and distributed architecture based on already existing protocols and specifications, such as UDP, SSDP, SOAP (Curbera et al. 2002) or XML (Bray et al. 2000).

In addition, it is supported by the Internet protocol family TCP/IP, which (independent of the company, operating system and programming language) enables the APIs of the devices connected to a network control to negotiate and exchange information and data in an easy and transparent way for the user. This way, the user does not need to be an expert in networks, devices or operating system configuration. In addition, UPnP technology does not depend on the physical environment, so it can work on the telephonic line, the power supply, Ethernet, radio frequency and IEE 1394.

1.2.2 General Features

The main characteristic of the protocol is the transparency of installing a device that has just been connected to the power supply. All the services of the installed device are automatically available without the need to configure anything in the protocol (Miller et al. 2001). UPnP notices when a new device is connected to the net, it gives it an IP address, a logic name and updates the rest of the devices about their functions and processing ability. As seems obvious, it also updates the new device about these same features of the others. This way, the user does not have to worry about the configuration of the net or losing any time installing new drivers or controllers for the devices. UPnP is dedicated to all these things each time a new device is connected (or disconnected) to (from) the net. It also optimizes the configuration of the devices.

Its application for development of a home automation system offers a new possibility to create distributed control architectures. In other words, robots have independent activation parts connected by an internal network. Because of this, UPnP gives more versatility and flexibility to the system. Moreover, any change in software or any device in the system is easily adaptable in the system.

A digital home based on UPnP is conceived to include all wire and wireless networks, entertainment devices, telephonic systems, home control and many more devices. It will also put some home networks together in a single logic made by programmable devices (Jeronimo 2004b) (Fig. 1.1).

One of the most common uses of this protocol is to enable devices or programs to open router ports, so they can properly communicate with the outside world.

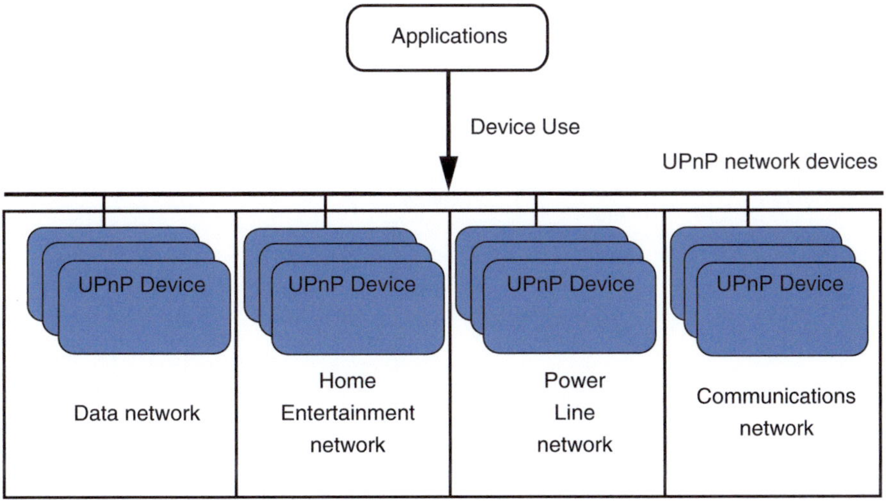

Fig. 1.1 UPnP: network unification technology (Jeronimo 2004)

1.2.3 Specific Features

Since the (Universal Plug and Play) UPnP model is based on the existence of two different components, the control point and the device, this protocol makes identification of the roles of every element in a home automation network possible. The main idea is that every device (a robot, a router, etc.) can be accessible on a local area network (LAN). Some will announce the services they offer to the rest using a protocol such as the SOAP or something similar.

An XML file with the name of the device and a description of the services that are offered are sent through the network each time a new device is plugged into the network. The file may also include a URL pointing to the website of the developer. In addition, an external pointer to detailed information about the services could be included.

This fact gives a clearer idea of the ease of maintenance and transparency of use that this architecture provides to applications and interfaces. As shown in the figure above, Dynamic Host Configuration Protocol (DHCP) servers and/or DNS may be available on the network, so a new device may automatically be configured on the network upon connection. The next step will be discovering services. To offer a better idea of the protocol's way of working, a basic scheme of the logic structure of a UPnP network is shown in Fig. 1.2.

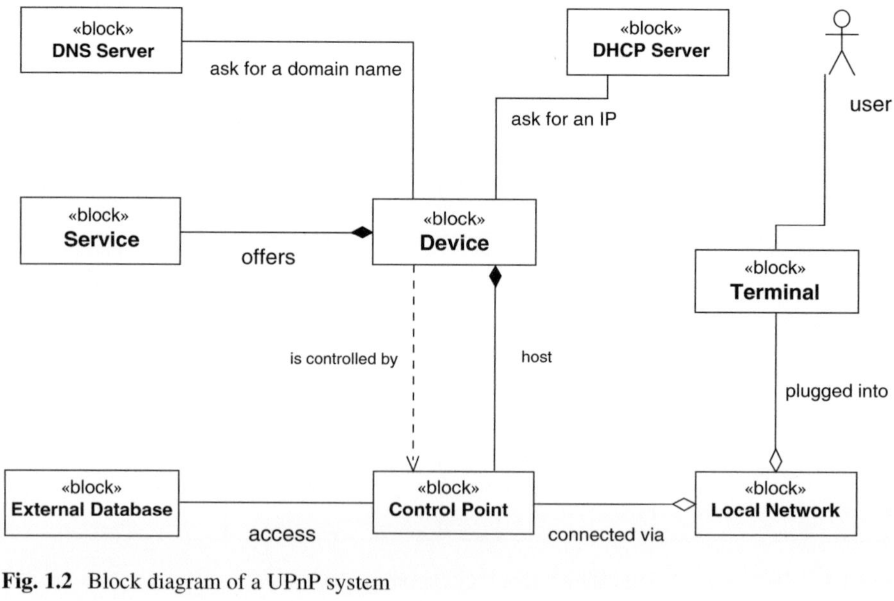

Fig. 1.2 Block diagram of a UPnP system

1.2.4 Protocols

1.2.4.1 TCP/IP

TCP/IP stands for Transfer Control Protocol/Internet Protocol. It is the grounding on which the development of other UPnP protocols takes place. TCP/IP is a set of protocols that covers different physical media and provides compatibility between different developers. It is based on the idea of an IP address or, in other words, the idea of providing an IP address to each computer connected to the network.

1.2.4.2 UDP/IP

The UDP (User Datagram Protocol) is the grounding that supports the HTTPU and HTTPMU sending of messages (see below). It makes the sending of datagrams possible before communication has been established.

1.2.4.3 HTTP, HTTPU and HTTPMU

These protocols are basic parts of UPnP. HTTP stands for Hypertext Transfer Protocol. HTTPU and HTTPMU are variants of HTTP, in particular, HTTP unicast and HTTP multicast. These variants are used for the delivery of messages over UDP/IP when multicast is used or it is not necessary to establish a connection (Fout 2001).

1.2.4.4 SSDP

The Simple Service Discovery Protocol (SSDP) is a protocol that allows searching for UPnP devices on a network. It detects devices and network services that use the SSDP, such as UPnP devices. It also detects SSDP devices and services running on the local computer. Searches are made by sending a SSDP search request (on HTTPMU). In addition, it can refine its search to find only devices of a particular type, only certain services or even a particular device. A message is sent to all the devices on the network through the same channel, so each device must be listening through the multicast port. When it receives a search request, it checks the search criteria and, if there is a coincidence, answers by sending a unicast SSDP message, on HTTPU, with the code "200 OK," which indicates that the request was successful. When a device is connected to the network, it sends several SSDP presence messages announcing the services it offers (delivery is not guaranteed over the UDP). The messages sent by the device have a link to the location of the document that contains its description, with its properties and the services it offers. In addition to the SSDP properties, it provides the device with methods for disconnection notification and updates the device's information using timeouts.

A SSDP packet is just an HTTP request with the statement “NOTIFY” (to announce) or with “M-SEARCH” (to find a service), leaving the HTTP body empty, and keeping UPnP-specific attributes in its head.

1.2.4.5 GENA

The General Event Notification Architecture (GENA) allows sending and receiving notifications using HTTP over TCP/IP and HTTPMU over UDP/IP. UDP multicast is useful because it allows a single report to be distributed to a potentially large group of receivers using a single request. GENA defines the terms of the subscriber and the publisher of the notifications, which enable the event’s mechanism used by UPnP to warn of changes in the state of services. When a subscription to a service takes place, it sends event messages updating the changes in the status of the device. These event messages are in XML format. Apart from this, GENA is also used to create presence messages, which are sent using the SSDP.

1.2.4.6 SOAP

The Simple Object Access Protocol (SOAP) provides a standard mechanism for packaging messages. It defines how two objects in different processes can communicate exchanging XML data. Thus, UPnP makes use of XML and HTTP to run remote procedure calls (RPCs), sending control messages to devices and getting the results or the errors in each case. Each control request is a SOAP message that contains the action invoked and all the necessary parameters. The response is another message of the same type with the state or the result of the action requested to the device.

Although many protocols are created to simplify the communication between applications (RPC from Sum, DCE from Microsoft, RMI from Java and ORPC from CORBA), the SOAP has received more attention because of the great support received from the industry. It has been accepted by almost all large companies. Consequently, it is becoming the standard for communication based on RPC over the Internet. Some of its advantages are:

- It is not associated with any language.
- It is not strongly associated with any transport.
- It is not tied to any distributed object infrastructure.
- It makes the most of the existing standards in the industry.
- It enables interoperability among multiple environments.

1.2.4.7 XML

Extensible Markup Language (XML) plays an important role in the exchange of data. It is similar to HTML, but its main function is to describe data and not to display them as is the case of HTML. XML is a format that allows reading data through different applications. Specifically, it can structure, store and exchange information (W3C 2008). It is used in UPnP for device and service descriptions, control messages and events.

1.2.5 Components of a UPnP Network

A UPnP network defines various types of components, such as control points, devices and services. These are detailed below.

1.2.5.1 Devices

UPnP devices are logical containers for a service or set of services, and sometimes for other devices (embedded devices). Embedded devices can be discovered and used independently of the main container. Each UPnP device can offer any number of services. By itself, a device merely provides a self-description of its information, such as developer, model name and serial number. Device services are those that provide real functionality (Fig. 1.3).

There are different categories of UPnP devices, standardized according to the set of services provided by each device. This information (along with properties such as those mentioned above) is saved in an XML document that must be kept in the device until it needs to be sent.

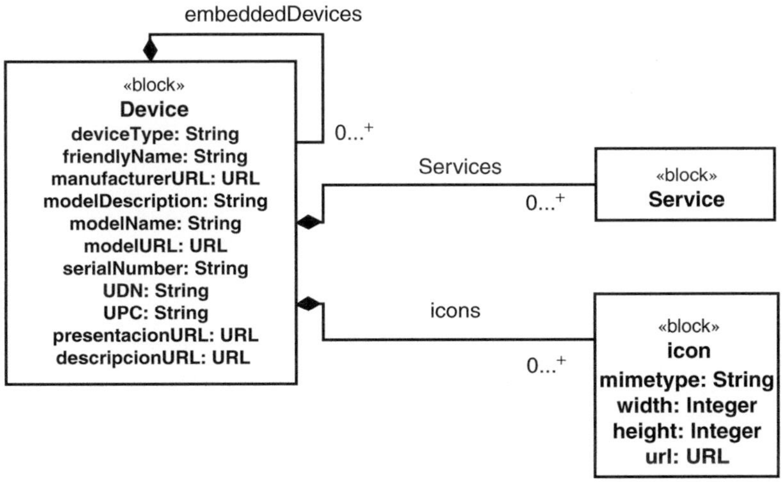

Fig. 1.3 (Unified Modelling Language) UML class diagram of a UPnP device (Jeronimo 2004)

1.2.5.2 Services

Each service in a UPnP device can contain any number of actions. An action has a name, a set of input parameters and a return value (optional). Each argument has a name, a value and an address. The address can be input or output depending on whether the argument is given to the action or returned by the action.

A service has a service identifier (URI) that identifies it from all the others; there cannot be two services with the same identifier. It can keep the variables that represent the current state of the service. These state variables have a name, type, default value, current value and a range of permissible values. If a variable sets an event to indicate a state, then it is an event notification variable.

A service is a state table, a control server and an event notification server. The state table contains the variables updated when there is any change in service status. The control server receives action requests and performs them, updates the state table and returns the result. The event notification server publishes updates of changes in the state of service (Fig. 1.4).

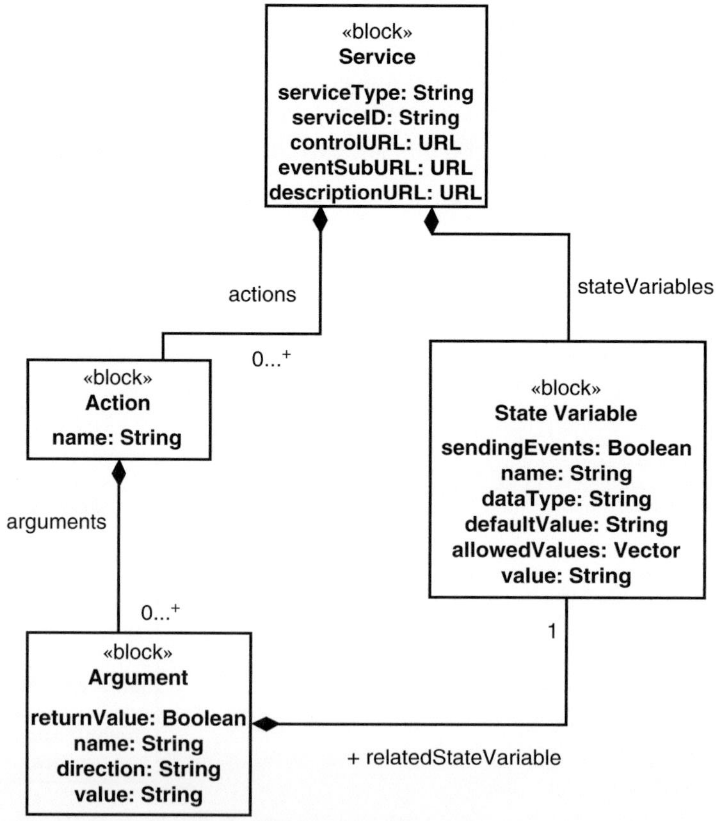

Fig. 1.4 UML class diagram of a UPnP service (Jeronimo 2004)

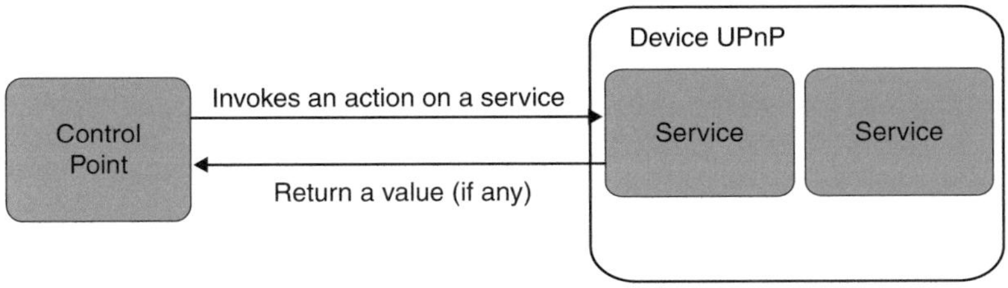

Fig. 1.5 Control point invoking a service action (Jeronimo 2004)

1.2.5.3 Control Points

A checkpoint is a network entity that invokes the functionality of a device. It is capable of discovering and controlling other devices. In terms of client/server in a UPnP network, the checkpoint will be the client and the server role will be played by the device.

Once the checkpoint finds the device, it is capable of:

- Getting the description of the device and a list of related services.
- Getting the descriptions of the services in which it is interested.
- Invoking actions to control the service.
- Subscribing itself to the service's events (Fig. 1.5).

When the status of the service changes, the event notification server sends an event to the checkpoint. In short, a checkpoint finds the devices, invokes the related actions to their services and signs up for event notifications. By contrast, a device responds to the actions invoked by the checkpoint and sends the events when the variables change their state.

1.2.6 UPnP Running

To describe the UPnP way of working, we will show the development in six basic steps or stages: Addressing, Discovery, Description, Control, Event Notification and Presentation. The routing stage can be considered step zero. The representation of the protocol stacks used in each one of the following steps is shown below (Fig. 1.6).

1.2.6.1 Addressing

Since all UPnP communications are based on the Internet Protocol (IP), a device must obtain an IP address before it can join to a network that supports UPnP.

The first step, also known as the ***zero phase***, is based on this; an address for the checkpoints and devices connected to the network must be obtained. All the reasoning presented in this phase is valid for both devices and checkpoints.

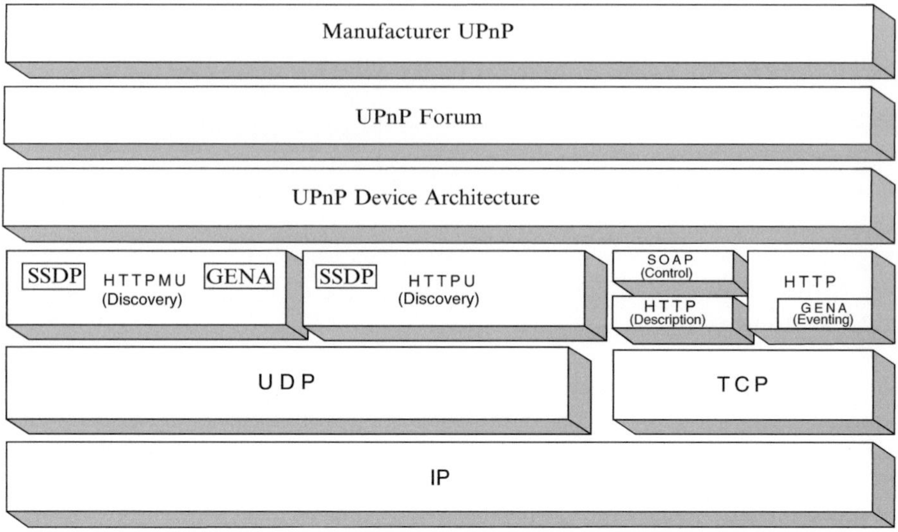

Fig. 1.6 Protocol stacks of UPnP running

Addressing is the process by which a device automatically gets an IP address. It allows a device to join to the network and be prepared for communication with other devices and checkpoints. The routing protocols implemented in the UPnP devices enables them to join dynamically to an IP network and to get an address without being configured by the user.

UPnP devices can use the DHCP, UDP-based, to retrieve an IP address from a DHCP server. To do this, both devices and checkpoints must have a DHCP client. Being connected to the network, the first thing to do is to find a DHCP server that provides them an address. If this server already exists on the network, they must use the address they have been assigned.

If the network does not have a DHCP server, automatic IP addressing (Auto-IP) must be used to get the IP address. Through this mechanism, the device takes a random address within the 169.254/16 range to minimize potential collisions with other devices. This range was established by the ICANN (Internet Corporation for Assigned Names and Numbers) and the IANA (Internet Assigned Numbers Authority) for IP self-configuration in private networks. Once assigned an IP address using Auto-IP, it must be checked that this address is not used by any other device on the network using the ARP (Address Resolution Protocol). Each device must periodically verify the existence of a DHCP server on the network to manage the process of addressing. In this case, the automatically assigned IP is ruled out and so they start with the dynamic address's assignment using the DHCP server.

First, a device or checkpoint tries to contact a DHCP server to obtain an IP address. If it is unable to locate the server, the device uses Auto-IP, which allows devices to select addresses without having a server to assign it to them. It may be necessary to resolve the assignment to IP addresses because the devices can implement protocol layers higher than UPnP. To obtain this functionality, devices must incorporate a DNS client and support the DNS dynamic registration.

1.2.6.2 Discovery

The discovery phase defines how a device announces its presence and how checkpoints discover it. A UPnP device is like a mini web server that can be detected and monitored by a checkpoint. The discovery process allows checkpoints to find devices and services of interest and obtain information about them. The devices use the SSDP to announce their services to checkpoints. These last ones use the SSDP to search for devices. In the tower of protocols below, you can see certain color codes that match the parts of each message defined below. These color codes are useful until the operation's description of the UPnP technology is finished (Fig. 1.7).

Once a device has acquired an IP address, the SSDP announces its services to all the control points of the network. Similarly, when you add a checkpoint to the network, the SSDP searches for relevant devices on the network. They will answer if there is an agreement with the data of the search message. The message exchanged in both cases is a discovery message that contains essential details about the device or its services, such as the type of device, identifier and a pointer to more detailed information (Fig. 1.8).

It must be kept in mind that when a checkpoint or device initializes and connects to the network, it must wait a random time between 300 and 3,000 ms before sending any message of discovery. These ranges are set to avoid problems when many devices connect to the network at the same time (300 ms) and to minimize delays in the recovery of a network (3,000 ms) (UPnP Forum 2001).

A URL of the XML document describing the device is included in the discovery and the announcement responses of every device. This URL provides the necessary

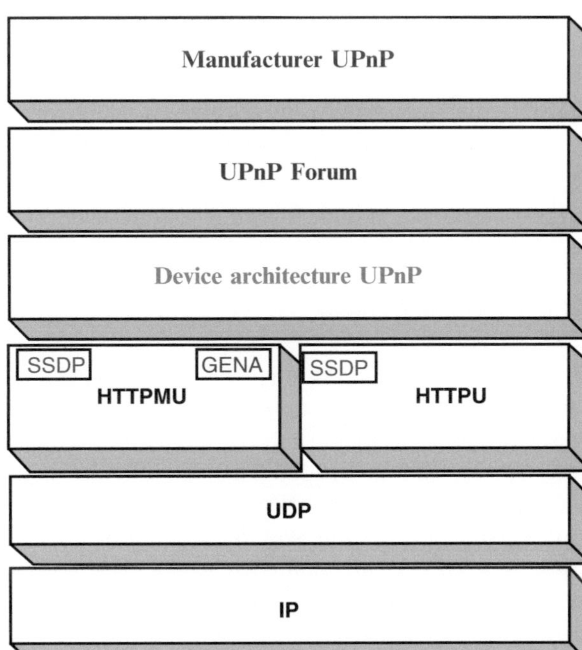

Fig. 1.7 Protocol stack for discovery

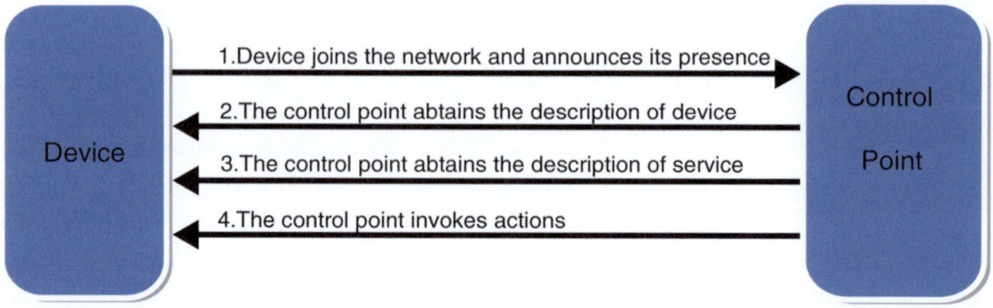

Fig. 1.8 Recovery of the descriptions of a service and a device (Jeronimo 2004)

information to the checkpoints to retrieve the descriptions of the devices and their services. All services contained in a device have three URLs that provide the necessary information to allow the checkpoints to communicate with them:

- The URL of **control** is where the checkpoint sends requests to control the service. UPnP device manufacturers specify one for each device.
- The URL of **subscriptions to events** is where checkpoints send requests to subscribe to events. There is a URL for this kind of service in each device. If a service does not have event variables, and therefore no notification of events, the element URL of subscriptions to events must appear, but it will be empty.
- The URL of **description** indicates the location of the checkpoints from which the service description document will be retrieved. The service description document is returned by an HTTP GET request.

A checkpoint has two possibilities to search for devices. It can pick up a notification message sent by a device or it can request the response of the device using a discovery message sent by the checkpoint itself (Fig. 1.9).

The devices must refresh their announcement messages periodically because they have a limited lifespan. For this reason, they are not obliged to cancel those sent previously (announcing their capabilities) when they disconnect from the network.

1.2.6.3 Announcement

Once a device joins to the network, it announces its embedded devices and its services to checkpoints through NOTIFY messages defined by GENA. These are multicast messages that use the SSDP. These messages are sent to the address and port (239.255.255.250:1900). This default value is indicated by ICANN/IANA to use it with the SSDP. The checkpoints are supposed to listen to arriving messages in this port, knowing this way the capabilities that are available on the network. These messages do not require a response. One important fact about announcement messages is the time of validity, which indicates the period in which the device is available. After finishing this period without sending an announcement message, the device will stop being available on the network.

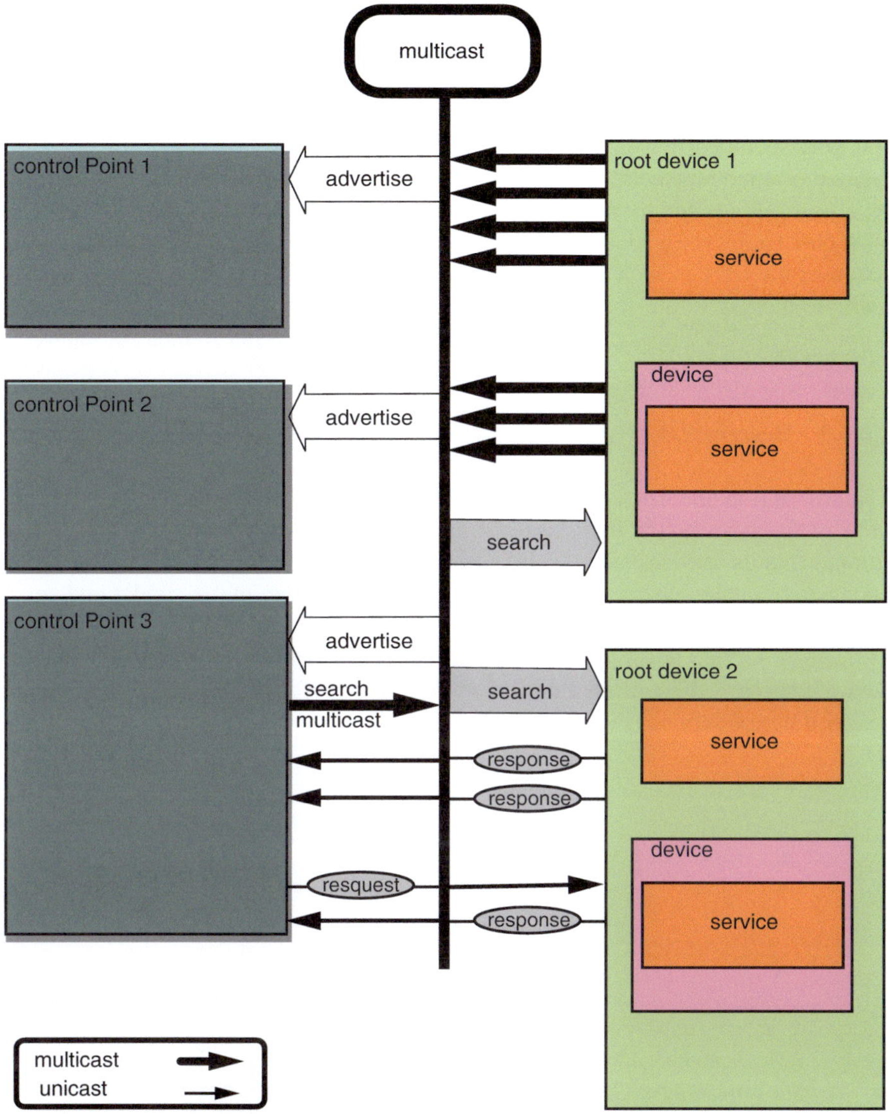

Fig. 1.9 Discovery (UPnP Forum 2008)

During the announcement process and considering that a root device has **d** embedded devices and provides **k** different types of services, a total of **3+2 d+k** announcement messages are sent to the network. This can be deduced, assuming they are different devices, by interpreting the number of messages that should be sent by a device:

- A message for each type of service with *NT=type of service*.
- A message for each type of device (root or embedded) with *NT=device type*.
- A message for each device (root or embedded) with *NT=UUID of the device*.
- A message regarding the root device with *NT=upnp*: *rootdevice*.

1.2.6.4 Search

This procedure is activated when a checkpoint requires a type of device or a specific service. This is when the control point sends a multicast message with the address and port specified above, i.e., 239.255.255.250:1900. In this case, unlike in the method of announcement, it will require answers from the devices that fit with the specifications defined by the checkpoint.

A checkpoint must send multiple ***M-SEARCH*** messages since the messages are sent over the UDP (which does not guarantee delivery). A control point will receive multiple messages, but some will be duplicates. To filter these replies, the control point uses the USN header, which provides a unique identifier to look for answers.

1.2.6.5 Description

The description enables a device to list all the features it can provide. The descriptions of the devices and their services are stored in XML documents. The device summarizes its services and capabilities in a standard format. A device description document contains device information (such as developer, make, model and serial number), a list of the services provided by the device and a list of its embedded devices. A service description document contains detailed information about the device's service, the actions that the service provides, the parameters and values returned by the service (Fig. 1.10).

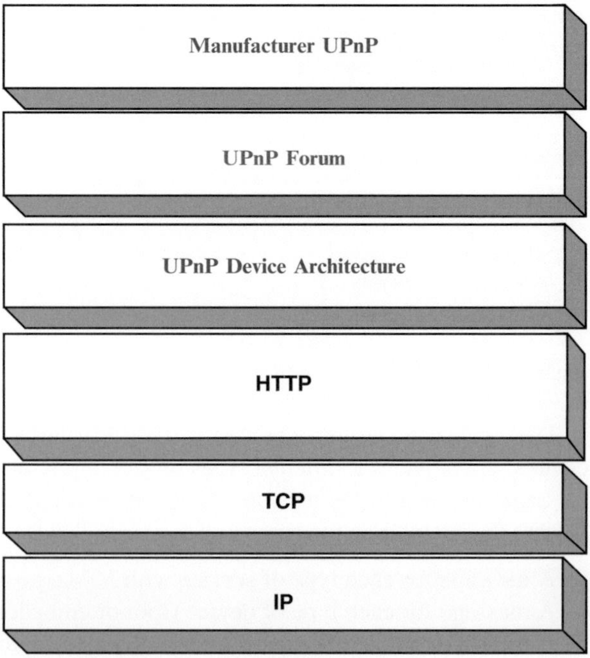

Fig. 1.10 Stack of protocols for description

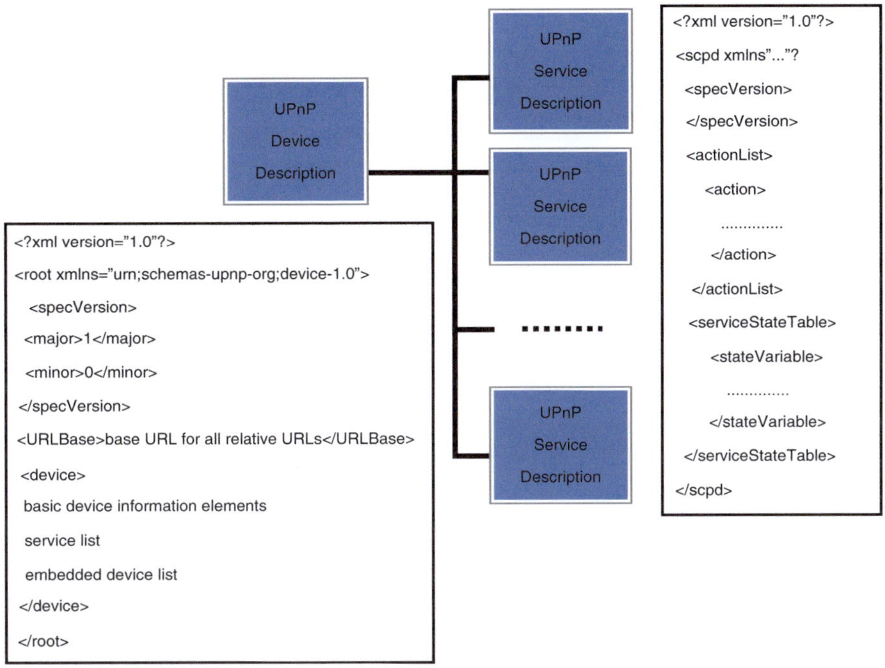

Fig. 1.11 Hierarchy in the device description and service (Jeronimo 2004)

The answers to the search messages received by a control point contain URLs that provide descriptions of the capabilities of the device. Control points use these description documents to get more information from the devices, trying this way to get their features and interact with them.

The description of a UPnP device consists of two parts: the device description, which refers to the physic and logic container, and the service description, which refers to the capabilities offered by the device. Both descriptions are provided by the developer and are written in XML.

Devices may contain other logic devices apart from services. The UPnP description document includes a list of integrated devices and a description of the available services. For each service, its description includes a list of actions to which the service replies and the arguments for each action. The service description also includes a list of variables that reflects the state of the device. These variables are described in terms of their types of data, ranges and characteristic events (Fig. 1.11).

To receive the description of a device, the control point sends an HTTP request using the GET method to the URL contained in the discovery message that had previously been received by the device. When it receives the request, it replies with an HTTP message that contains the device's description in the message's body.

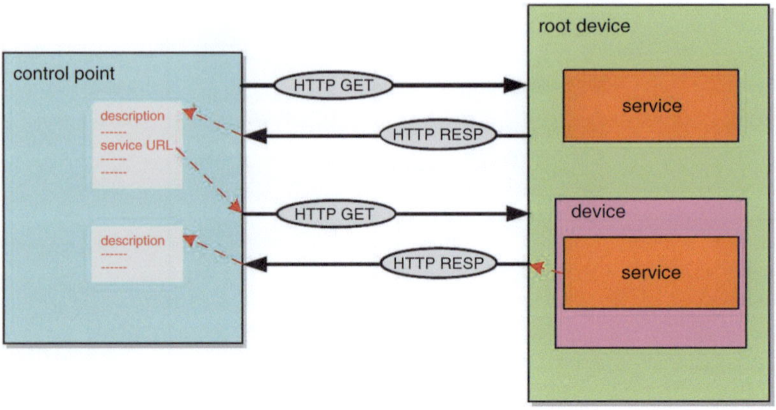

Fig. 1.12 Description (UPnP Forum 2008)

The URLs of the device's description of its services are included in this description. The information contained in the device description consists of:

- An XML document containing various data from the device.
- The meanings of all nested devices.
- A list of all services supported by the device, including state variables and actions.

The control point can send another HTTP request containing the URLs of the service descriptions to reacquire the service descriptions. The format of the control point's request is shown below (it is important not to forget about the blank line at the end of the header):

GET: description route HTTP/1.1
HOST: host:port
ACCEPT-LANGUAGE: control point's favorite language.

The syntax of the device's response message is shown below, and the device's or service's description will appear in the body (Fig. 1.12).

HTTP/1.1 200 OK
CONTENT-LANGUAGE: language used in the description
CONTENT-LENGTH: length of the body, in bytes CONTENT-TYPE: text/xml
DATE: time to answer

For each service that contains a device, the description contains (in addition to what was stated above) the name and type of service, service description URL, URL for control and URL for event notification. Finally, the device description also provides a description of all nested devices and a URL for presentation (Fig. 1.13).

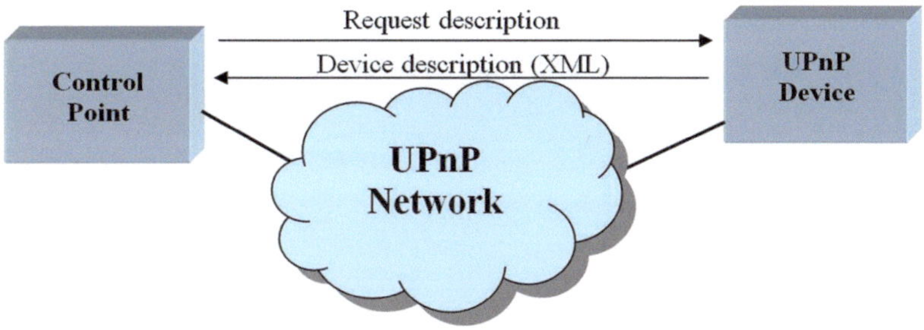

Fig. 1.13 Request/reply of the description and its protocols

Fig. 1.14 Protocol stack to control

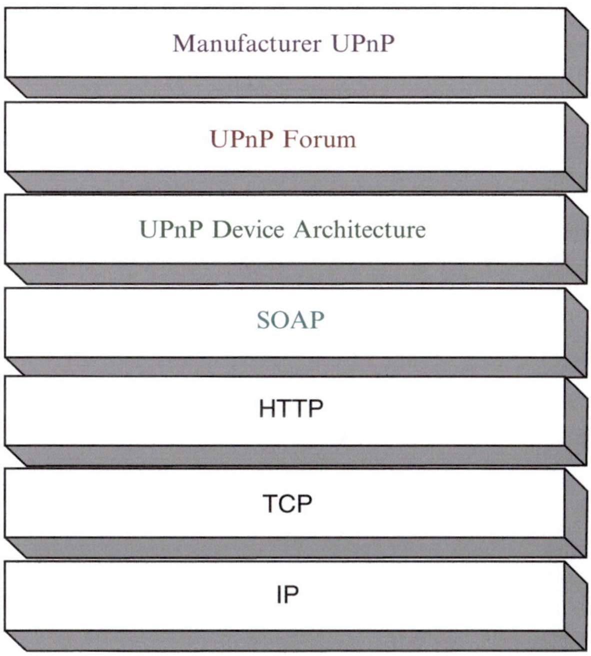

1.2.6.6 Control

Control is the UPnP phase in which the control points invoke actions to the services of the devices. Once a control point has all the information about a device and one of its services through its description, it will be able to control the service by invoking actions. The protocol stack that supports the control phase in the running of UPnP is shown in Fig. 1.4.

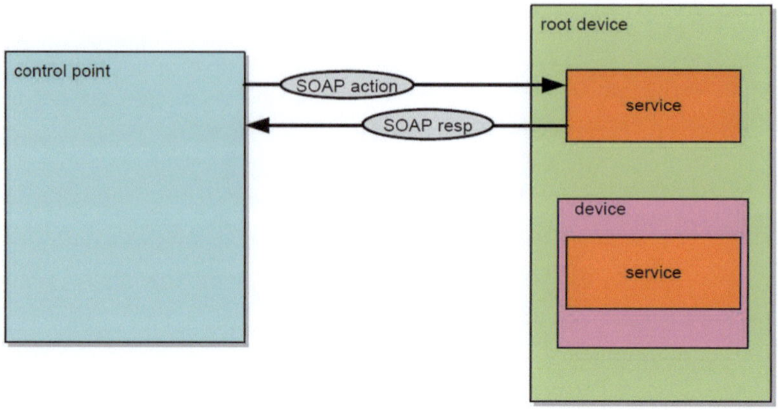

Fig. 1.15 Control (UPnP Forum 2008)

To control the device, UPnP is based on the SOAP, which uses XML and HTTP to provide web messaging and RPC. XML makes public the content of the message and HTTP sends the message to its destination. The SOAP comprises four parts:

- *Extensible and required envelope to encapsulate the data.* The SOAP envelope defines a SOAP message, and this is the basic unit of exchange between SOAP message processors. This is the only obligatory part of the specification.
- *Optional rules* for encoding data represent data types defined by the application.
- *Link between SOAP and HTTP*. This part is also optional since the SOAP can be used in combination with any transport protocol or mechanism that can transport the SOAP envelope.

RPC Model. Its purpose is message exchange (request/response). It is a convention to represent RPC and its responses (Fig. 1.14).

To invoke an action, the control point sends a message to the control URL that it already knows from the description phase explained above. The device will respond with the result or the errors obtained after running the service action. Moreover, this action may change the state of the service and, therefore, change some of its variables (Fig. 1.15).

To invoke a specific action, the control point must send a SOAP request using the POST method to the service device. This control message contains information specific to the manufacturer, name of the action, names of the arguments and variables that are defined by the UPnP Forum.

Requests for the state variables were considered in UPnP, but this way of working has been discarded by the UPnP Forum and must not be used for control points. Instead, the working committees and the manufacturers define actions that return the variable's value and that can be invoked by the control points.

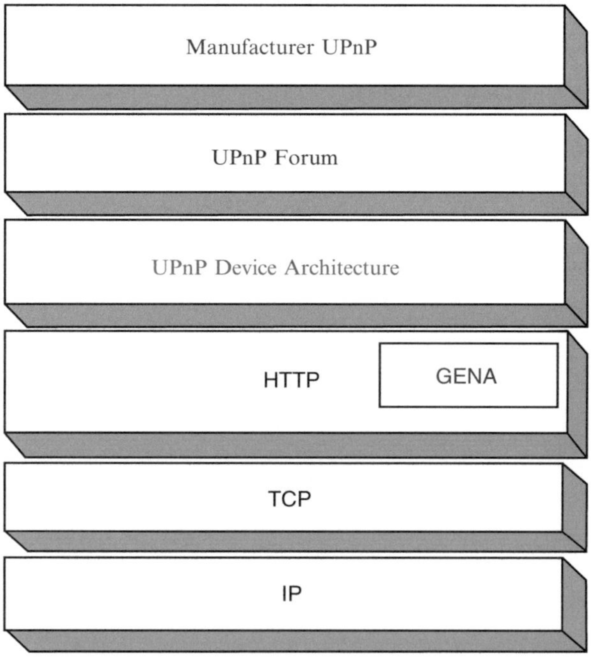

Fig. 1.16 Stack of protocols for the notification of events

1.2.6.7 Event Notification

Event notification offers the possibility of notifying a control point when the state of a device changes. As explained above, a service description contains a list of variables that model the state of the service. If any of these variables is likely to be reported as an event, the service publishes updates when any of these variables are modified. The protocol stack used in this case is shown in Fig. 1.16.

The event notification system uses a publisher/subscriber model in which the control points can subscribe to events sent by a service. The services publish event notifications to subscribers. An event is a message sent from a service to the subscribed control points. The events inform the subscribed control points about the state changes in the service.

A control point that wants to be notified about changes in the state of the variables subscribes to an event source by sending a subscription request to the URL of the events, which is contained in the corresponding device description. The subscription application must include the service to subscribe, a URL to send events and a subscription time.

If a service accepts the subscription request, it responds with a unique identifier of subscription (SID) and the life of the subscription, which indicates its validity period. This unique identifier allows the control point to refer to the subscription service for future applications to the service, such as renewing or canceling the subscription (Figs. 1.17 and 1.18).

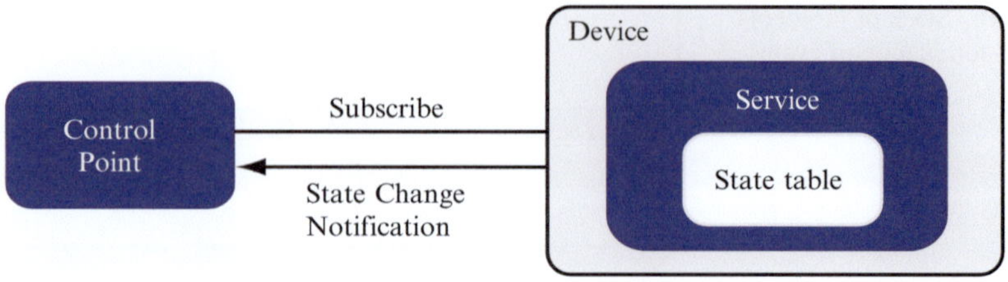

Fig. 1.17 Subscribing and notifying (Jeronimo 2004)

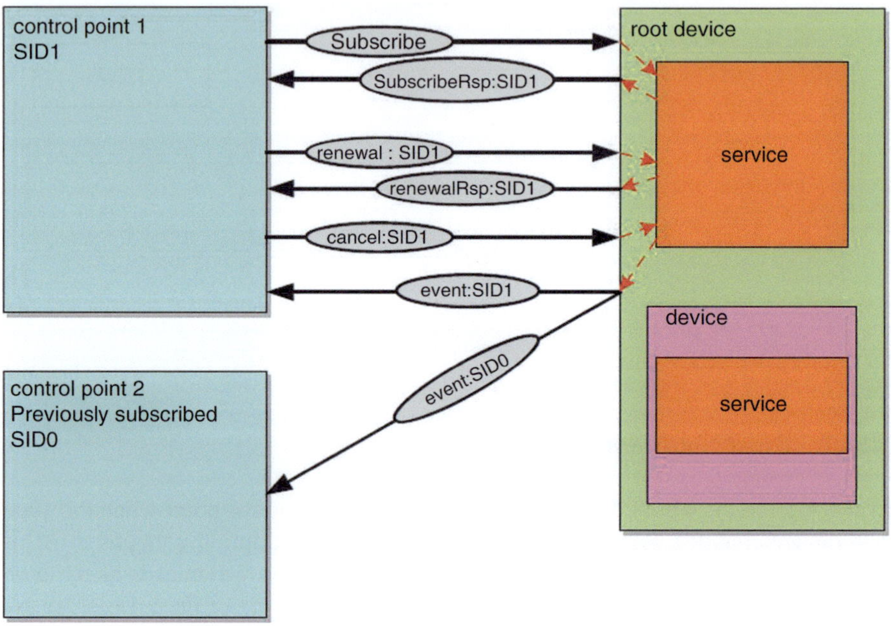

Fig. 1.18 Event notification phase (UPnP Forum 2008)

Event messages are sent to all subscribers regardless of the reason for the change in state variables. These messages contain information expressed in XML with the names and values of those state variables configured in the service as event variables. The event notification protocol is GENA and, as seen in the previous protocol stack, it is used in the TCP transport, which guarantees message delivery to the subscriber. When the subscription expires, the subscription identifier becomes invalid and the service stops sending events to the corresponding control point. If this control point attempts to send any message (renewal or cancellation, but not the subscription), the service is rejected because the ID is no longer valid.

The control point will send a subscription message to the URL of the service to receive its events. This message uses the SUBSCRIBE method defined by GENA and its syntax is:

SUBSCRIBE event route of the service HTTP/1.1
HOST: host:port

CALLBACK: <delivery URL>NT: upnp:event
TIMEOUT: request for the lifetime, in seconds

A blank line must be added at the end of the last header. When this message is received, the service establishes a list of subscribers with the following information for each of them: SID, URL for the event messages delivery, event counter and length of subscription.

If the subscription is accepted, the service sends a message with the identifier of the subscription and validity period. This message has the following syntax. It is important not to forget about the final blank line:

HTTP/1.1 200 OK DATE: request time
SERVER: OS/version UPnP/1.0 product/version
SID: uuid: subscription UUID
TIMEOUT: lifetime of the subscription, in seconds

The first event notification message must be sent after the message above. It contains the names of the variables and their current values in XML. In addition, each time that one of these variables, which are set as event variables, changes the service, it must send an event message to all subscribed control points.

These event notification messages are labeled with a different key for each subscriber to detect errors. In every control point, in the initial event message, this key is set at zero and increases with each subsequent notification message. This way, if the subscriber receives a notification with an incorrect key, it will reply to the service with an error message.

All subscriptions must be renewed periodically for the control points to go on receiving notifications. To keep a subscription active, the control point must send a renewal message before the subscription expires. The renewal message is sent to the same URL as the original subscription message, but this time it does not include the URL for event message delivery. Instead, the renewal message includes the subscription identifier received in the initial message, which confirmed the subscription. We can see this message format below and, as already mentioned, it must include the blank line:

SUBSCRIBE: service route HTTP/1.1
HOST: host:service port
SID: uuid:susbcription UUID
TIMEOUT: request for the time of subscription, in seconds

The answer to this message is exactly the same as in the subscription message case. When a control point does not want to get any more events from a service, it can call off its subscription by sending a cancellation message:

UNSUBSCRIBE: service route HTTP/1.1
HOST: host:service port
SID: uuid:subscription UUID

The answer to this message is, as in the case above, an HTTP confirmation. If the control point abruptly disconnects from the network without sending the message to

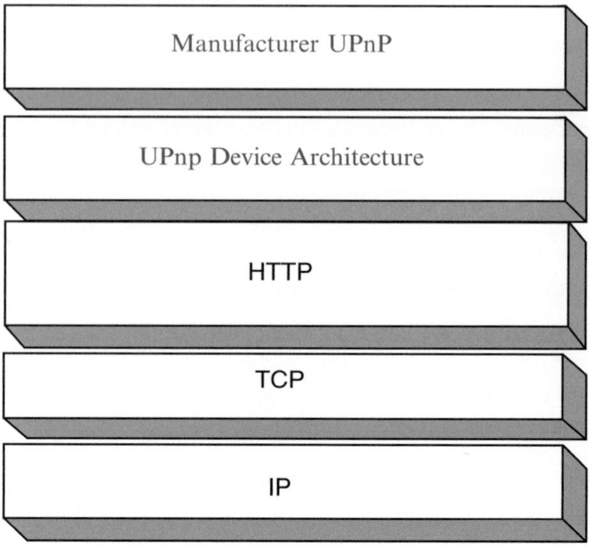

Fig. 1.19 Protocol stack for presentation

cancel the subscription, the service will keep on sending it notifications until the subscription time expires.

1.2.6.8 Presentation

In a UPnP network, a control point can monitor a device or check its status through the presentation of an HTML page. A home page can be loaded by the control point in a browser and this allows users to view and control the device. The protocol stack required for this is shown in Fig. 1.19.

Home pages are not necessary; if a device has no home page, it can still be controlled through standard control messages. If the device allows a home page, its description document contains the URL for the presentation page on the label *<presentationURL>*. This label must always be present. If the device has no home page, the label will be empty.

In the presentation phase, the control point sends an HTTP request using the GET method to the presentation URL (available in the device description) and the device then returns to the home page. After loading the page in the browser, the control point can monitor the device or check its variables. The diagram below shows this (Fig. 1.20).

The presentation message for requests includes the field *ACCEPT-LANGUAGE,* and the language of the presentation page will be defined by the *Content-Language* field, which is defined in the device. Figure 1.21 shows the way to recover a presentation interface from a device and the protocols used to do it.

An additional component of a UPnP network is the *application layer*. The capabilities of a device are defined by itself and the service models that provide the framework for the network components (description, control and events). A device manufacturer can develop these models by itself or work with other manufacturers

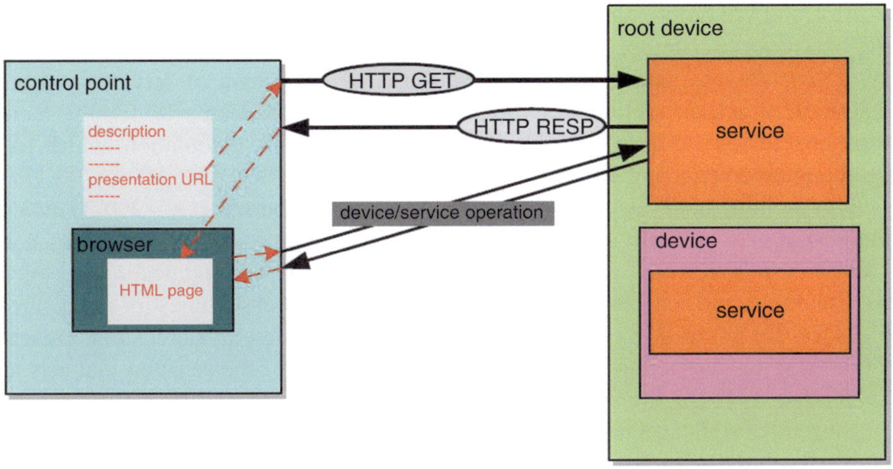

Fig. 1.20 Presentation (UPnP Forum 2008)

Fig. 1.21 Request/reply of a presentation interface and its protocols

inside the UPnP Forum to prepare the standard for the device and the service models. Currently, the working committee for UPnP has developed definitions for standard models.

1.2.7 Penetration in the Market

Some modern projects work with UPnP, such as Ahn et al. (2005), which compares it with other distributed systems such as CORBA. Kim et al. (2002) used UPnP to build a middleware layer for a home network. It is important to notice that, for this project, we have used device emulators (TV, fridge, etc.). This will be important for future implementations of the final solution.

Mok and Wu (2006) demonstrated UPnP protocol integration in a system consisting of a robot to manipulate objects. This paper described the design of the system systematically, the data used to compile the XML document of services and the definition of actions and control variables. In Maestre and Camacho (2009), there were conclusions to develop a flexible and low cost home automation, which has been implemented using UPnP.

This architecture is also used for sensor networks. Dobrescu et al. (2007) studied the design of a network with an interface between wireless sensor networks and UPnP via TCP/IP. This application makes possible the communication between control points and sensors and provides the use of web technologies for the control interface. By contrast, Song et al. (2005) discussed the few resources that present the sensing devices using a UPnP agent (BOSS, bridge of the sensors) that acts as an interface between the PC and the elements not supported by UPnP.

Currently, you can find different solutions in the market for developing UPnP systems, highlighting initiatives such as CyberLink for Java (Satoshi 2004), a Java implementation that automatically controls all the internal aspects of the protocol and allows the programmer to focus on the business layers and the tool's interface.

1.3 OSGi

1.3.1 Introduction

OSGi [Open Services Gateway Initiative (OSGi Alliance 2003)] is an independent corporation that brings together about 40 companies in an alliance responsible for defining and promoting open specifications for the delivery of managed services in network environments. It is based on the modularity of the Java environment, trying to abstract the implementation of components (bundles) using services to communicate. One of its main aims was to resolve certain development and deployment conflicts, such as class conflict and the explicit dependencies (Bartlett 2009).

1.3.2 General Features

OSGi is based on a layer model (Fig. 1.22) that includes, among others, the bundles or packages (components developed as jar files), services (which provide communication between bundles through Java objects), modules and security layer.

The features of OSGi could be a good alternative for the development of complex systems because of its versatility and cross-platform feature (only a JVM resident on each node of the network would be required for the running).

Some examples of the use of OSGi in network systems can be found; for example, Gu et al. (2004) demonstrated the use of an intelligent system (SOCAM)

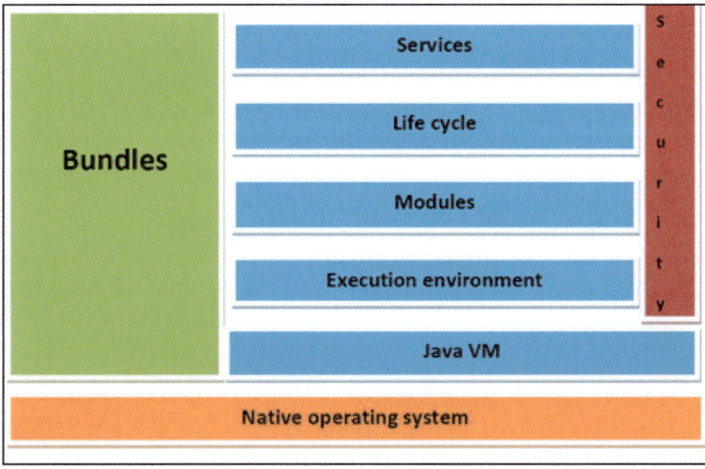

Fig. 1.22 Layer model of an OSGi system

based on the ontological model that was aware of the running context. Other examples include Kang et al. (2005), whose way of working is based on UPnP AV (a UPnP variant for multimedia devices) for the multimedia internal service, and in OSGi, which treats each entity of the UPnP system as a "bundle" of services. Nevertheless, OSGi has several problems (some delimited, some not) because of its poor basis of compatibility and poor management of dependencies.

1.3.3 Specific Features

1.3.3.1 Framework of OSGi

The OSGi framework is a module system for Java that implements a complete and dynamic model of components, which does not exist in independent environments of JVMs. The applications and components (which come in packets or bundles) can be installed, started, stopped, updated and uninstalled remotely without rebooting. The management of Java packages and classes is carefully specified. Lifecycle management is performed through APIs that make possible the remote download of management policies. The registry allows service bundles to detect if services have been added or deleted and acts accordingly.

Originally, it was focused on service gateways, but the scope has since widened. OSGi specifications are now used in applications ranging from cell phones to the Eclipse development environment (open source). Other application areas include automotive, automation in industry and buildings, PDAs, grid computing, entertainment (such as iPronto), fleet management and application servers. Figure 1.2 shows the hierarchical structure of an OSGi system.

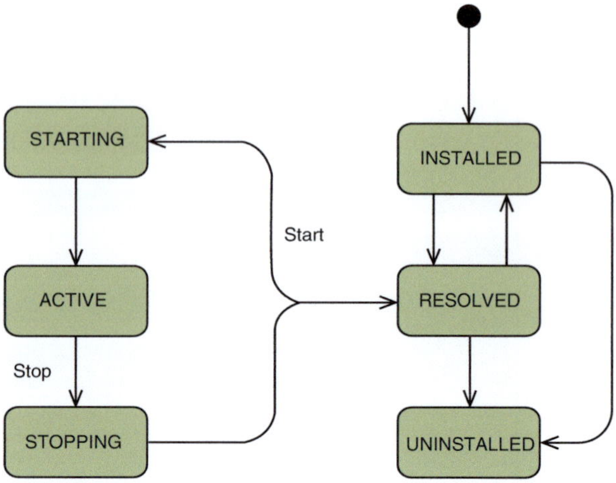

Fig. 1.23 Lifetime of an OSGi bundle

1.3.3.2 Specification Process

OSGi specification has been developed by its members in an open process that is available for the public free of charge under the OSGi specification license. The OSGi Alliance has a performance program that is open to its members. In September 2008, the list of certified OSGi implementations contained five entries.

1.3.3.3 Architecture

Any framework that implements the OSGi standard provides an environment for the modularization of applications in small packages. Each package is a collection of well-coupled and dynamically loadable classes, jar files and a configuration that explicitly state their external dependencies (if any). The framework is conceptually divided into the following areas (see Fig. 1.23):

- ***Bundles***: These are jar components with extra headers in a detailed manifest file.
- ***Services***: The service layer connects bundles dynamically, offering a model of publication, search and link to plain old Java objects.
- ***Register services:*** The API of some management services (ServiceRegistration, ServiceTracker and ServiceReference).
- ***Lifetime***: The API for the management of lifetime (install, start, stop, update and uninstall) bundles.
- ***Module***: The layer that defines the encapsulation and declaration of dependencies (how a bundle can import and export code).
- ***Security***: The layer that deals with security issues, limiting the functionality of the bundles to predefined capabilities.

- ***Running environment***: This defines what methods and classes are available on a specific platform. Since they are susceptible to change, there is no fixed list of running environments. The Java community is creating new versions and editions of Java constantly.

1. Bundles
   A bundle is a set of Java classes and additional resources accompanied by a detailed manifest file (MANIFEST.MF) of all its contents as well as the additional services required to provide the included group of Java classes more complex behavior until the point of abstraction, where the whole is treated as one component. An example of a MANIFEST.MF file, typical of OSGi headers, is shown below:

   Bundle-Name: Hello World
   Bundle-Symbolic Name: org.wikipedia.helloworld Bundle-Description: A Hello World bundle Bundle-ManifestVersion: 2
   Bundle-Version: 1.0.0
   Bundle-Activator: org.wikipedia.Activator
   Export-Package: org.wikipedia.helloworld;version = "1.0.0" Import-Package: org.osgi.framework;version = "1.3.0"
2. Lifecycle
   A lifecycle layer adds bundles that can be installed, started, stopped, updated and uninstalled dynamically. The bundles trust in the module layer for class loading but they add an API to manage the runtime modules. The lifecycle layer provides mechanisms that are not usually part of an application. Some extensible dependency mechanisms are used to ensure the correct working of the environment. Lifecycle operations are fully protected by the security architecture (Table 1.1).
3. Services
   The OSGi Alliance has specified many services, all of them by a Java interface. The bundles can implement this interface and register it with the service registry. Service clients can find it in the service registry or detect it when it appears or disappears (Tables 1.2–1.4).

Table 1.1 Description of the lifetime of an OSGi bundle

State of the bundle	Description
Installed	The packet has been successfully installed
Resolved	Every Java class that needs the bundle is available. This state indicates that the packet is ready to be started or stopped
Starting	The package is being started; it will call the method BundleActivator. start, and this one has not finished yet. When the bundle has an activation policy, it will remain in the initial state until it is activated, according to this policy
Active	The package has been successfully activated and it is running. Its starting method, Bundle Activator, has been called and it has returned
Stopping	The packet is being stopped. The method BundleActivator.stop has been called but the stop method has not returned yet
Uninstalled	The packet has been uninstalled. It cannot be changed to a different state

Table 1.2 OSGi system services

System services	Description
Logging	The information register, warnings, debugging and errors are handled through this service. It receives log entries and dispatches others bundles that have already subscribed to this information
Configuration admin	This service allows an administrator to set and view information about the configuration of the bundles
Device access	This simplifies the detection and connection of existing devices. It is used in Plug and Play environments
User admin	This service uses a database containing user information (both public and private) to issue authentication and authorization
IO connector	This service is implemented in the packet CDC (http://en.wikipedia.org/wiki/Connected_Device_Configuration)/CLDC (http://en.wikipedia.org/wiki/CLDC) javax.microedition.io (http://java.sun.com/javame/reference/apis/jsr118/javax/microedition/io/package-summary.html) as a service. This one allows the bundles to provide new protocol diagrams
Preferences	It offers an alternative, friendlier mechanism with OSGi to use the default Java package java.util. Properties (http://java.sun.com/javase/6/docs/api/java/util/Properties.html) for storage preferences
Component runtime	The dynamic nature of the services – they can be opened and folded at any time – makes it difficult to write software. Runtime component specification can make it easier to manage these issues, providing a declaration of XML-based units
Deployment admin	This standardizes the access to some responsibilities of the administration agent
Event admin	This provides the bundle with a mechanism of internal communication, based on a publish and subscribe model
Application admin	This simplifies the management of an environment with different kinds of applications that are simultaneously available

Table 1.3 OSGi protocol services

Protocol services	Description
HTTP service	This allows the information to be sent and received by OSGi using HTTP
UPnP device service	This specifies how OSGi bundles can be developed to work with devices UPnP (http://en.wikipedia.org/wiki/Universal_Plug_and_Play)
DMT admin	This defines an API to deal with a device using concepts of the specifications for device administration from Open Mobile Alliance (http://en.wikipedia.org./wiki/Open_Mobile_Alliance) (OMA)

Table 1.4 OSGi miscellaneous services

Other services	Description
Wire Admin	This allows the connection between producers and consumers
XML parser	This service allows a bundle to locate a parser (XML syntax analyzer) with specified properties and compatibility with JAXP (http://en.wikipedia.org/wiki/JAXP)
Measurement and state	This allows and simplifies the correct use of measurements in an OSGi platform of service

1.3.4 Organization

The OSGi Alliance was founded by Ericsson, IBM, Motorola, Sun Microsystems and others in March 1999 (before becoming a nonprofit corporation called Connected Alliance).

Among its members (as of May 2007) are more than 35 companies from different business fields, such as IONA Technologies, Ericsson, Deutsche Telekom, IBM, Makewave – before it was Gatespace Telematics – Motorola, Nokia, NTT, Oracle, ProSyst, Red Hat, Samsung Electronics, Siemens, SpringSource and Telefónica.

The alliance has a board that establishes the governance of the organization. OSGi officers have different roles and responsibilities to support the alliance. The technical work is carried out in the expert groups (EGs) organized by the board of directors, and the non-technical work is carried out in various working groups and committees. The technical work in EGs includes development specifications, reference implementations and compliance testing. These EGs have made four versions of OSGi specifications (as at 2007).

There are EGs dedicated to business areas, mobile phones, vehicles and central platforms. The Expert Group Company is the latest EG and handles applications regarding the company/server. In November 2007, the Residential Expert Group began working on specifications to remotely administer residential gateways or homes.

1.3.5 Penetration in the Market

In October 2003, Nokia, Motorola, IBM, ProSyst and other members of OSGi formed the Mobile Expert Group, which specifies a service platform based on MIDP for the next generation of smartphones, dealing with some of the needs that CLDC cannot handle. MEG joined OSGi at the same time as R4.

Also in 2003, Eclipse selected OSGi as the runtime platform for the plug-in architecture to be used for the Eclipse Rich Client Platform and the IDE platform. Eclipse itself includes sophisticated tools to develop OSGi bundles, and there are some plug-ins for Eclipse to improve the development of OSGi (for example, ProSyst and Knopflerfish have Eclipse plug-ins available for OSGi developers).

There is a free software community with activity around OSGi. Some open source implementations are widely used such as Equinox OSGi, Apache Felix, the OSGi Knopflerfish project and the editing of embedded server Equinox (mBedded Server Equinox Edition, BSEE). Now talking about the support to the system's development and testing, projects Pax OPS4J provide a lot of components and useful knowledge.

1.4 Jini

1.4.1 Introduction

Jini is an architecture that provides an infrastructure for defining, publishing and search services on a network. It was developed with Java classes (Arnold 1999). The main feature of Jini is the service discovery in multicast mode or search mode for specific services (similar to the idea of UPnP). It uses the multi-platform feature of the Java platform to provide universal services, registering each one as serialized objects with their own interfaces. A diagram of Jini's architecture is shown in Fig. 1.24.

The main aims of Jini's platform are the immediate availability of services, the hardware abstraction on the Java environment, the service-based architecture and simplicity.

1.4.2 General Features

This is an easy protocol (Morgan 2000; Fig. 1.25). When a device connects, it registers in the lookup service of the Jini network. After that, the service sends a file with the bytecode that a customer needs to use its services.

- The lookup service stores this file in a special table and puts similar services together in groups. When a client asks the search service to use a device, it responds with a list of devices that provide these services.
- The client responds with the identifier of the specific device to be used and the search service responds with the bytecode mentioned above.
- The client will now be able to use the bytecode (during a specific time, in a shared way or in an exclusive one).

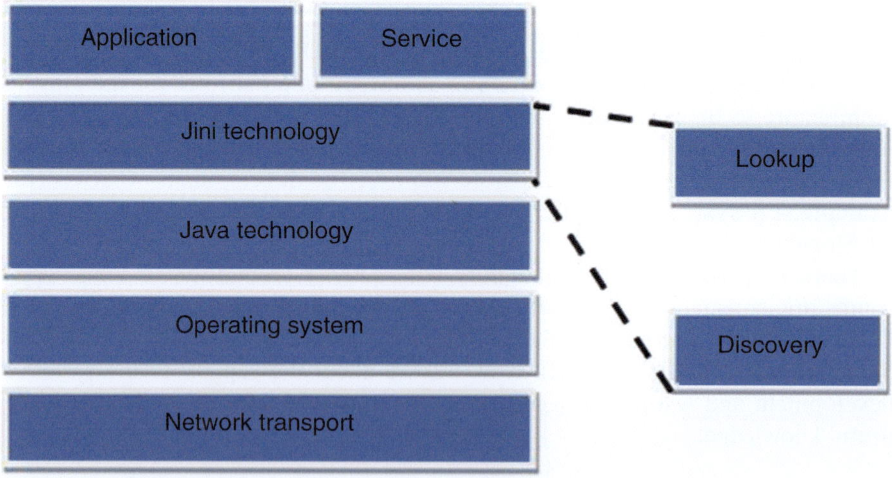

Fig. 1.24 Layers model of a Jini system (Allegro 2006; Gupta et al. 2002)

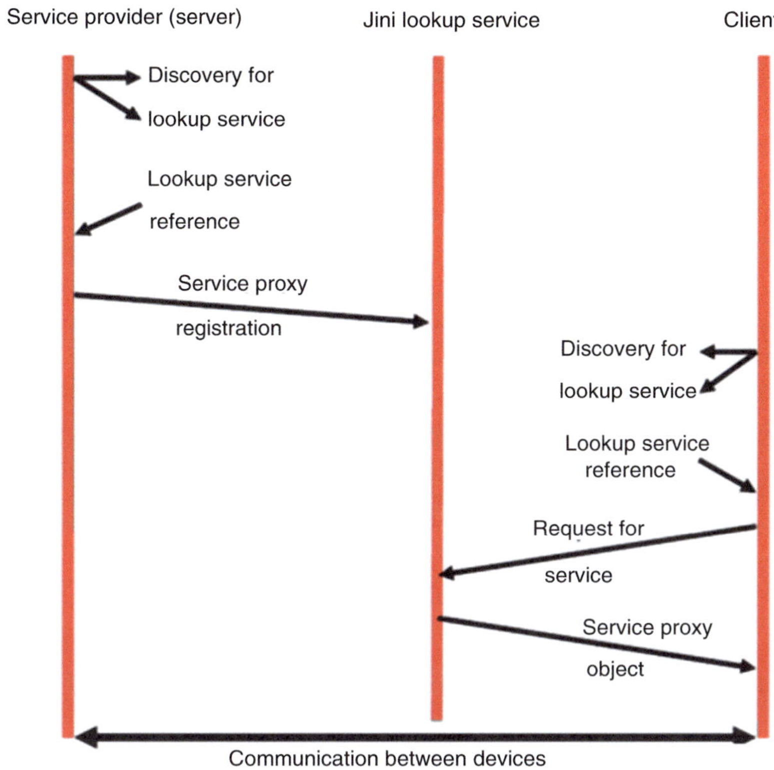

Fig. 1.25 Diagram of the sequence events of Jini (Allegro 2006; Gupta et al. 2002)

1.4.3 Specific Features

The purpose of the Jini architecture is to put devices and software into groups inside a distributed and dynamic system. This simplifies the access, management and maintenance services offered by each point separately, keeping the flexibility and control offered by a personal computer.

1.4.3.1 Services

The most important concept within the Jini architecture is the service. A service is an entity that can be used by one person, one program or another service. It may be a calculation, saved data, a communication channel with another user, a software filter, a hardware device or another user. As an example, we can mention the printing services of documents.

Members of a Jini system share access to services. A Jini system should not be considered a set of clients and servers, users and programs or even programs and files. Rather, it consists of a set of services used to perform a particular task.

The services may use other services, and the customer of a service can be a service itself for other customers. The dynamic nature of a Jini system enables services to be added or removed, at any time, from a set, according to demand, need or the changing demands of the working group. Jini systems provide mechanisms for service construction, lookup, transfer and use in a distributed system. The services may be:

- *Devices*, such as printers, screens and discs.
- *Software*, such as applications or utilities.
- *Information*, such as access to databases or/and files.
- *Users* of the system.

Services communicate with each other using a service protocol (set of interfaces written in Java). All these protocols are undefined. The groundings of the Jini system define a small number of these protocols, which in turn define the interactions among critical services.

1.4.3.2 Lookup Service

Services are found and resolved by a lookup service. This service is the central mechanism for the system to boot and the main point of contact between the system and users. In other words, it is a mapping service made up of lookup interfaces that indicates the functionality provided by a service to groups of objects that implement the lookup service. In addition, descriptive entries associated with a service allow finer lookup services, based on properties understandable by a human being.

The content of a lookup service may include other search services, providing hierarchical searches. In addition, this kind of service may contain objects that encapsulate other names or service directories, providing a system of pointers that connects Jini lookup services with other search services. Thus, references to a Jini lookup service can be mixed with these names and directory services, providing the customers of these services with a way to access a Jini system.

A service is added to a lookup service by a pair of protocols called discovery and join. First, the service locates an appropriate lookup service (using the discovery protocol) and then joins (using the protocol join).

1.4.3.3 Java Remote Method Invocation (RMI)

Communication between services can be performed using Java RMI. This infrastructure is not a service itself, but is rather part of the Jini technology infrastructure. RMI provides mechanisms to locate, activate and perform the garbage collection of Java objects.

RMI is a Java extension to traditional mechanisms for RPC. RMI not only allows data to pass from one object to another through the network, but also whole objects can be sent and received, including their codes. Much of the simplicity of the Jini system is because of this ability to move code through the network, encapsulated in an object.

1.4.3.4 Security

The design of the Jini security model is based on the concepts of a master list and an access control list. Jini services are accessed by an entity – the principal – which generally refers to a particular user in the system. The services themselves may request access to other services, providing the identifier of the object that implements the service. The access of an object to a service depends on the content of an access control list associated with the object.

1.4.3.5 Leasing

Access to many of the services in the Jini system environment is based on the concept of lease or loan. A loan is a grant of guaranteed access to a service for a certain period of time. Each loan contract is negotiated between the service user and provider, as part of the protocol service: a service is requested for a certain period of time and access is granted for the same time period (probably taking into account the time span taken to make the application). If a contract is not renewed before it is released – because the resource is no longer necessary, the client or the network fails or the contract cannot be renewed – both the user and the resource provider may agree that the resource can be released.

Leases are exclusive or non-exclusive. The first ensures that no one else can have a contract on the resource during the contracted period, whereas non-exclusive leases permit multiple users to share the same resource.

1.4.3.6 Transactions

A series of transactions, in a single service or spanning multiple services, may be involved in a transaction. The Jini transaction interfaces provide the service protocol needed to coordinate a two-phase commit. The responsibility for deciding how to implement transactions – and the semantics in a transaction – is left to the services themselves using these interfaces.

1.4.3.7 Events

Jini supports distributed events. An object may allow other objects to register in the events of an object and receive a notification with their histories. This allows event-distributed programs to be written with a great variety of liability and scalability guarantees.

Table 1.5 Components of RMI

	Infrastructure	Programming model	Services
Basic Java	JVM	API Java	JNDI
	RMI	JavaBeans	Enterprise Beans
	Java security	…	JTS
Java + Jini	Discovery/Join	Leasing	Printing
	Distributed security	Transaction	Transaction manager
	Lookup	Events	JavaSpaces services

1.4.3.8 General View of Components

The components of a Jini system can be divided into three categories: infrastructure, model of programming and services. The infrastructure is the set of components that builds a Jini system, whereas services are the entities inside it. The programming model is a set of interfaces that allows the construction of reliable services, including those that are a part of the infrastructure and those that are a part of the whole.

These three categories, although disjunct, are intertwined in a way that makes distinctions between them confusing. It is also possible to build systems with some of the features of the Jini system with variants on the categories or without any of them. By contrast, the main feature of Jini is that it is a system built with a particular infrastructure and described programming models, based on the concept of service.

The separation of the segments in the architecture means that only a slight change is needed in the inherited code to be used in a Jini system. However, the power of a Jini system is only available for services built using the integrated model from the beginning. A Jini system can be viewed as an extension of the network's infrastructure, programming model and services that made Java technologies popular in the case of a single machine. These categories, along with the components for the Java application environment, are shown in the table below (Table 1.5).

1.4.4 Organization of the Jini Architecture

1.4.4.1 Infrastructure

The infrastructure defines the basic core of this technology. It includes:

- ***A distributed system for security***, integrated in the RMI, which extends the security model from Java to the world of the distributed systems.
- ***The discovery and join protocols***, service protocols that allow other services (hardware or software) to discover, be a part of and announce the services offered to the other members of the group.
- ***The lookup service***, which is used as a backup for the services. The entries in the lookup service of objects are written in Java, and they can be downloaded as a part of a search operation and work as local proxies for the service that sets the code in the lookup service.

1.4.4.2 Programming Model

The infrastructure enables the programming model and makes use of it at the same time. The contracts made in the lookup service have a limited lifetime. This fact allows the lookup service to precisely check the set of available services at a specific moment. When the services binds to or separates from the lookup service, the events are notified about it, and the objects that have already shown an interest in receiving this information are updated about these new or defunct services.

The programming model is based on the ability to move the code, supported by the infrastructure. Both the infrastructure and the services that use it are calculation entities that live in the physical environment of the Jini system. However, services also constitute a set of interfaces that define the communication protocols used by services and the infrastructure to communicate between them.

These interfaces together form the distributed extension of the standard model in Java programming, which constitutes the Jini programming model. Among the interfaces that make up the Jini model are:

- *The leasing interface*, which defines a way to allocate and release resources through a model based on the renovation of their lifetime.
- *The event and notification interfaces*, which are extensions of the event model used by JavaBeans components for distributed environments. This feature allows event-based communication between services enabled by the Jini technology.
- *Operation interfaces*, which enable entities to cooperate so that all changes occur in the group or none take place.

1.4.4.3 Services

The technology infrastructure and the Jini programming model are designed to enable the services to offer themselves and to be found on the network. These services make use of the infrastructure to call and discover each other and announce their presence to other services and users.

The services appear as objects written in Java, perhaps made up of other objects. A service has an interface that defines the operations that may be requested of it. Some of these interfaces are intended to be used by programs, whereas others are intended to be administered by the client to enable the service to interact with a user.

The kind of service determines the interfaces of which it is composed and defines the set of methods used to access the service. A service can be implemented only by other services. Some of the Jini services are:

- *A printing service*, which can print from Java applications.
- *A service of JavaSpaces*, which can be used for simple communication and storing groups of objects written in Java.
- *A transaction administrator*, which allows groups of objects to participate in the Jini transaction protocol defined by the model of programming.

1.4.5 Penetration in the Market

There have been various initiatives to implement Jini as a form of communication between devices. In this regard, we underline the Ronin Agent Framework (Chen 2000), an environment based on Jini's distributed agents. This implementation attempted to improve the initial protocol, making it independent of the domain (so external devices could communicate with the local network), among other advances.

Furmento et al. (2002, 2004) described the implementation of a SOA [architecture oriented to services (He 2003)], ICENI, using the Jini platform, among others. This environment is based on an independent specification of SOA.

The integration of different service platforms is not easy. Allard et al. (2003) integrated Jini with UPnP, but it is important to note that these two protocols are incompatible by themselves. This new platform allows UPnP services to use Jini devices and vice versa by making just a few configuration changes. However, these authors do not answer several of the questions referring to the limitations encountered when working with these two architectures together.

In addition, for devices that do not have enough capacity to run a Java Virtual Machine (JVM), Jini offers the possibility of using a surrogate host. This is just another device capable of supporting a JVM, which works as a bridge between the original device and the Jini network architecture.

1.5 DLNA

1.5.1 Introduction

The DLNA (Digital Living Network Alliance) is an international and collaborative organization of companies involved in consumer electronics, industrial computers and mobile devices.

DLNA is a standard that allows different devices from the same network connected together to share information easily and without complicated configurations (Fuentes 2007). This system works with both wireless and Ethernet networks and even with the power supply. The DLNA has established a set of standards for the platforms and infrastructure software to be completely compatible. It focuses on the interoperability among mobile devices associated with multimedia images, digital audio and digital video.

Thus, assuming that all available devices on the network support this technology, a copy of the content and the network can be accessed from any device. In other words, we can listen to music from the files stored on our computer, watch movies stored on the digital video recorder on our computer or see photos of our camera on the TV. Figure 1.26 shows a possible scenario using this technology.

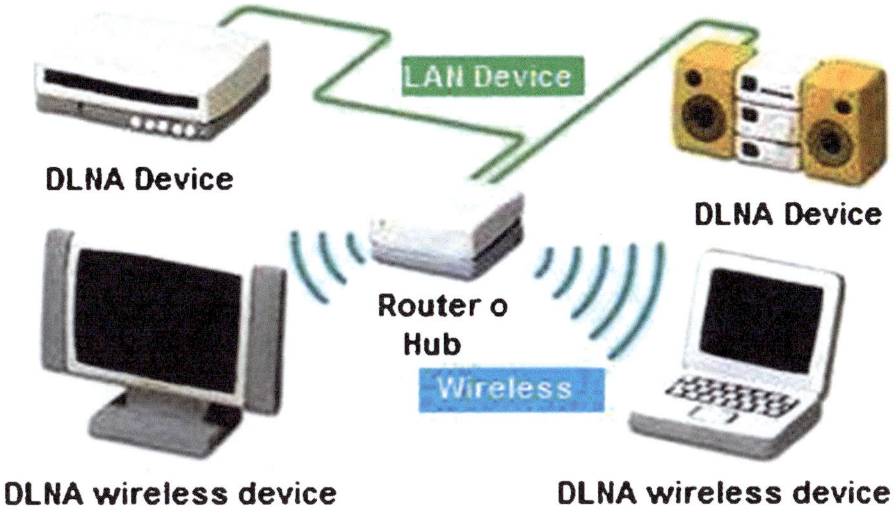

Fig. 1.26 Interoperability between two devices using DLNA

The objectives proposed by this technology are listed below:

- Digital music should be easily captured, stored and accessed from anywhere in the house. Digital photos should be managed, viewed and printed very easily.
- It must be possible to read content anywhere and enjoy it while traveling by car or walking down the street (there are already projects to synchronize information).
- It must be possible to save the distributed content to be able to see it as many times as we want.

1.5.2 General Features

The digital home is an electronic network made up of PC and mobile devices that cooperate transparently. The aim of DLNA is to become a home network for all its global customers. This objective integrates the interoperability of the three digital islands within the home: the Internet, broadband electronic network and island of mobile devices (Fig. 1.27).

The DLNA network must have at least a server and a client to work. The main objective of DMS (Digital Media Servers) is to provide multimedia content to DMP (Digital Media Players), which act as clients. These devices include camcorders, digital cameras, game consoles and mobile phones, but they need to be certified, that is, they must have integrated the electronics and configure to the DLNA standard.

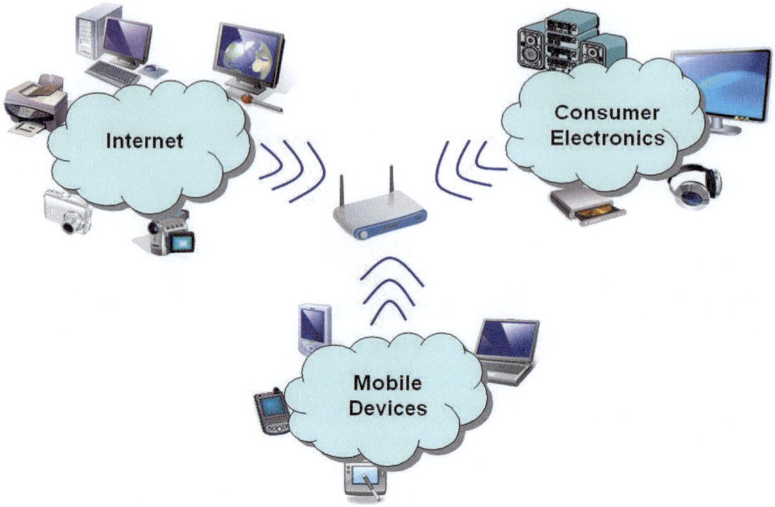

Fig. 1.27 Objective of DLNA: digital islands at home

DLNA makes use of a part of the technology developed for UPnP that allows the discovery of other devices on the local network. DLNA is based on UPnP and IETF (Internet Engineering Task Force) technologies (DLNA 2007). The DLNA standard is based on standards established in the industry and developed by groups such as IETF, World Wide Web Consortium (W3C), Motion Picture Experts Group (MPEG) and the UPnP Forum. Interoperability between devices is transparently performed by providing a particular service to the user. This includes the ability of the devices to communicate with each other and exchange useful information.

The interoperability guidelines require that all devices must support connectivity via Ethernet, Wi-Fi or Bluetooth. It uses TCP/IP for all network connections and works with HTML and the SOAP for transport and media management. The required formats to support images, audio and video are also defined. They are JPEG, LPCM (Linear Pulse Code Modulation) and MPEG2, respectively.

DLNA is based on a specification created by the working groups of the UPnP Forum. This specification is the UPnP AV (Audio and Video UPnP), and it has been the greatest success for these working groups, at least in terms of digital content (Fig. 1.28).

1.5.2.1 DLNA Model for Devices

The model for devices used by DLNA comes from the UPnP Forum and consists of devices, services and control points. The devices are network entities that provide services. These services are the basic control units and they perform actions to keep a state through its variables. The control points are network entities used to discover

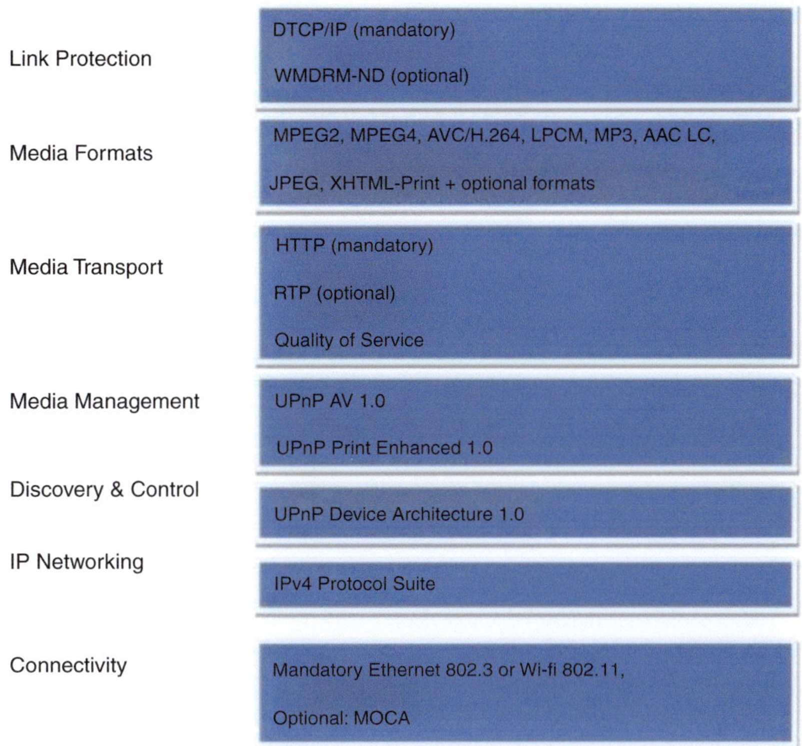

Fig. 1.28 DLNA interoperability model

and control other devices on the network. A group of multiple devices can be controlled by a control point.

In the UPnP standard, interoperability was first between the control point and a single device. However, with the evolution of the specification of UPnP AV (and DLNA as well) the basic model of devices was improved. For this reason, although interoperability between the control point and device still works, it has been extended to other devices so that they can interact with each other by exchanging digital content using different communication protocols (Fig. 1.29).

There are 12 kinds of DLNA devices in three different categories.

The category Home Network Device (HND) consists of five classes of devices that share the same use on the network system, with the same media formats and connectivity requirements.

- *DMS*. These are devices that can originate, acquire, record and store media on the model of interoperability in the digital home. There are DMS that help to protect the content saved. These devices, in case a customer is not able to handle a particular format, must be able to convert the file into another format before sending. Some examples of these devices include digital video recorders,

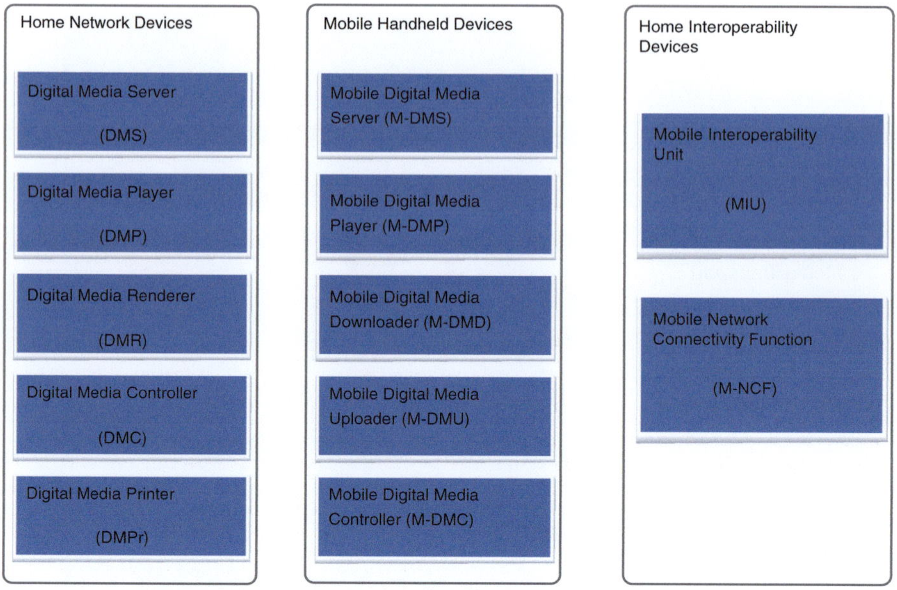

Fig. 1.29 Categories and kinds of DLNA devices

computers, home cinema with hard disk drives (such as music servers), devices to capture video and images and multimedia mobile phones. We can see the protocols and services of DMS in Fig. 1.30.

- *DMP.* These devices select and play the digital media stored on the network and include TV monitors, home cinema, PDAs, multimedia mobile phones, consoles and digital media adapters.
- *Digital Media Renderer (DMR).* Devices that reproduce the content received from DMS or their mobile counterparts after being configured by another device on the HND, such as a digital media controller (DMC, see below). DMC and mobile DMC devices will be explained in subsequent studies. Examples of such devices include televisions, audio/video receivers and remote speakers for music. The services and protocols of a DMR are show in Fig. 1.31.
- *DMC.* This device has the ability to find content exposed by DMS and adapt it to the rendering capabilities of a DMR, establishing the connections between them. It can also send instructions to another device, such as telling a server to play a particular video on a TV or sending a photo to a printer. A possible example of a DMC could be a learning remote control or a multifunctional device such as a multimedia mobile phone.
- *Digital Media Printer (DMPr).* These devices provide printing services to the home network. Some examples are a network printer or an application running on a PC with a USB-connected printer.

The category Mobile Handheld Device (MHD) consists of five classes of devices that use the same model as in the HND category, but have different requirements for

Fig. 1.30 Protocols and services of DMS

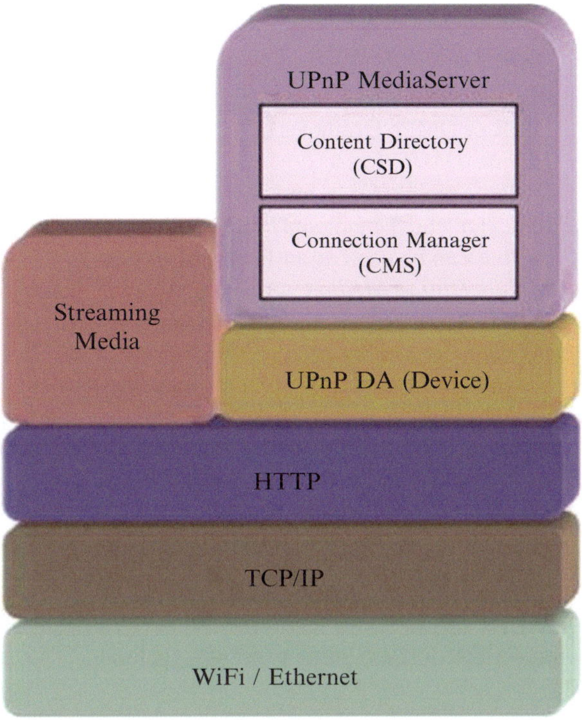

Fig. 1.31 Services and protocols of a DMR

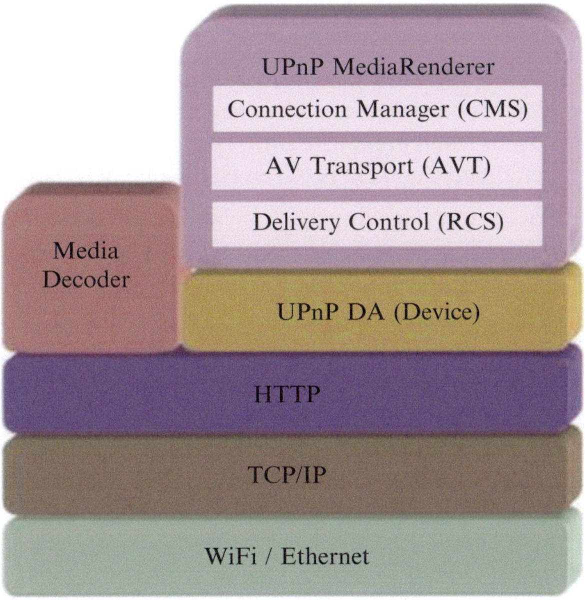

media formats and network connectivity. This category includes the following kinds of devices and features:

- ***Mobile DMS***. Wireless devices that provide and distribute content to a mobile DMP, DMR or DMPr. Examples of these devices are mobile phones and music players.
- ***Mobile DMP***. These devices are able to find and play the content offered by DMS or mobile DMS and play it in a local environment. An example of this kind of device may be a media tablet, which is a portable player with Wi-Fi connectivity that can be used as an Internet browser.
- ***Mobile Digital Media Controller***. A device that finds content offered by a mobile DMS and adapts it to the capabilities of a DMR, establishing connections between the server (DMS) and renderer (DMR). A PDA and an intelligent remote control are examples of such devices.
- ***Mobile Digital Media Uploader***. These wireless devices send (load) a mobile DMS or DMS with an upload functionality. A digital camera and a phone with an integrated camera are examples of such devices.
- ***Mobile Digital Media Downloader***. This finds and downloads the content exposed by DMS or mobile DMS and reproduces it after downloading. An example is a portable music player.

MHDs interact with stationary devices in the DLNA digital home and allow a wide variety of uses. Some examples include:

- Play images and videos taken from a MHD on a TV.
- Remote control function.
- Uploading images, music and video clips to a media server.
- Download images to a server using its controls.

The category Home Infrastructure Device (HID) integrates two kinds of devices. These devices are designed to enable MHDs and HNDs to interact.

- ***Mobile Network Connectivity Function***. These devices provide a bridge function between the network connectivity of MHDs and HNDs.
- ***Media Interoperability Unit***. Devices that make possible the change of format in multimedia content between HNDs and MHDs.

These 12 kinds of devices enable the sharing of digital content over a network. The three basic classes that must be in a DLNA network are DMS, DMP and DMC, and a particular device can do the functions of one or more of these basic devices.

The ways of working of these devices on the DLNA network, or the phases that it has to carry out, are similar to those described for UPnP. In the next picture, we see a representation of how to proceed in DLNA (Fig. 1.32).

1.5.3 Specific Features

Nowadays, the IPv4 protocol family is used, but the IETF is standardizing IPv6 as an enhancement of this version. The use of IP in the digital home brings us many benefits:

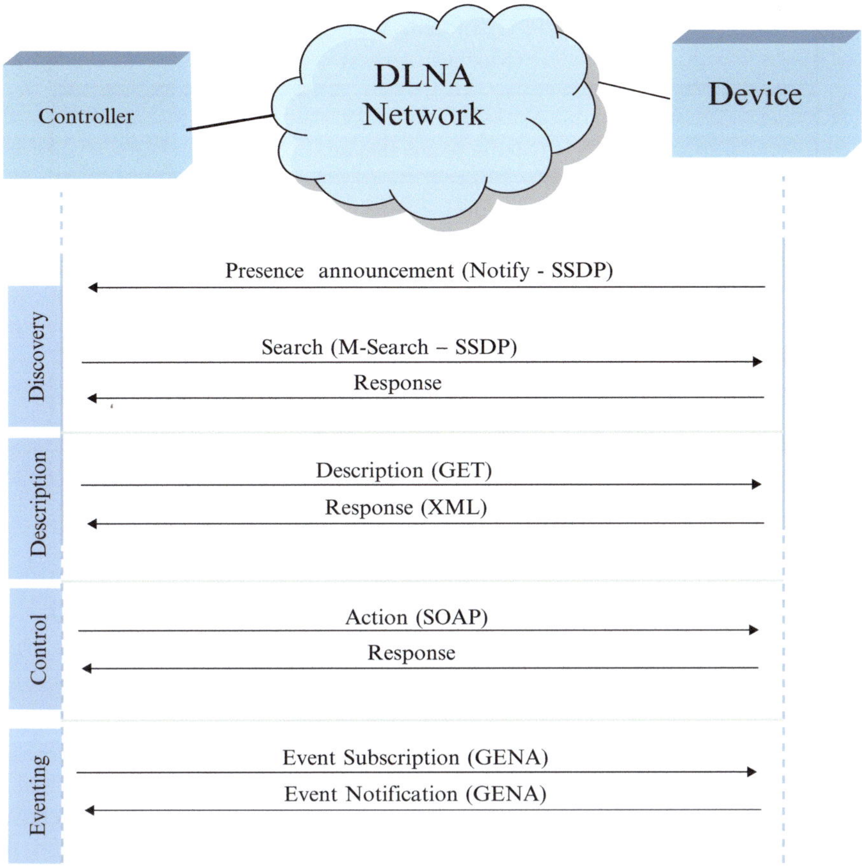

Fig. 1.32 DLNA working (Heredia 2008)

- It allows us to run applications over different means that can communicate in a transparent way. IP provides the framework that allows applications to be independent of the transport technology.
- It allows connecting all the devices in the home to the Internet. Using IP, every digital home device can connect to any other connected to the Internet.
- IP connectivity is cheap. Its implementation makes sure that IP is available at a lower cost than that of other technologies.

Therefore, IP support in the current digital home is essential for interoperability among devices. The graphic below shows the protocol stack used by DLNA 1.5 (Fig. 1.33).

The base for DLNA is the TCP/IPv4 protocol. Each device must implement a DHCP client and look for a DHCP server the first time it connects to the network. The device must use the IP address assigned by this server and, in case it does not find any server, the device will use Auto-IP, which means that it will generate an IP address

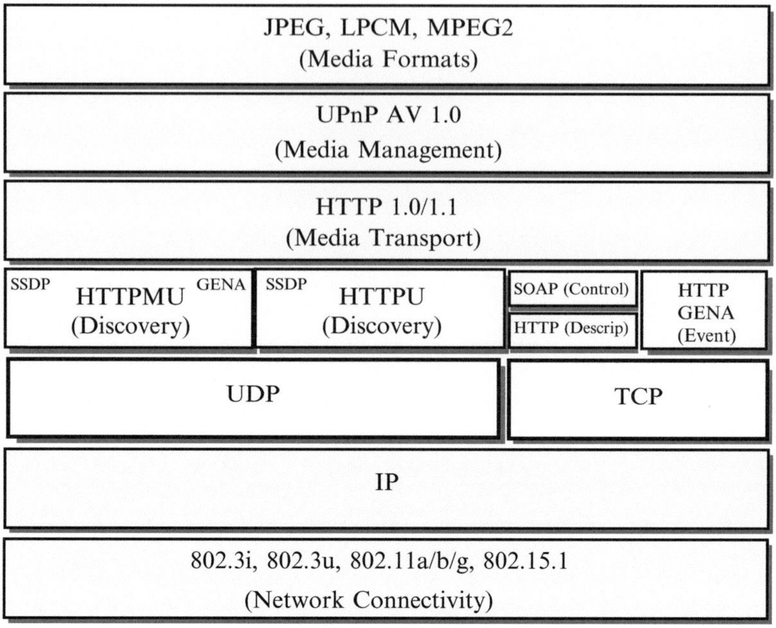

Fig. 1.33 DLNA stack of protocols

within the 169.254/16 address range. The first and last 256 addresses in this range are reserved and cannot be used. Once it has an address, it must determine whether that IP is available using ARP. If the device receives a response, it is assumed that the chosen IP is currently in use on the network and must generate a new one. In addition, the device must periodically check the existence of a DHCP server.

The technologies for the network connectivity that can be used in DLNA are Ethernet 10Base-T and 100Base-T (802.3i/802.3u) for wire connections, Wi-Fi (802.11a/802.11b/802.11g) for wireless connections and Bluetooth for wireless connections in handheld devices. In future, the idea is to start working with Ethernet 1000Base-T (802.3ab) and faster Wi-Fi connections (802.11n). It is also important to know that technologies such as LonWorks, CeBus, X-10 and Universal Powerline Bus are supported through UPnP bridges.

To protect digital media devices, DLNA technology makes use of digital rights management, which restricts the use of the media and devices. To protect the links (encryption/decryption) it is necessary to include a layer above all others in the protocol stack. This layer is based on Digital Transmission Content Protection (DTCP)/IP, which is needed to establish secure interoperability, and WMDRM-ND, which is optional and provides access to additional content.

DTCP/IP is a technology to protect links and is particularly adapted to work over IP (Arruda 2008). It is used to provide security to commercial content. It allows for the establishment of a secure authenticated channel that supports data flow (streaming) with limited copying rights: copy once, never copy and copy-restricted rights.

Table 1.6 Mobile household appliances

Media format	Mandatory formats for household devices	Optional formats for household devices	Mandatory formats for mobile devices	Optional formats for mobile devices
Image	JPEG	GIF, TIFF, PNG	JPEG	GIF, TIFF, PNG
Audio	LPCM (2 channels)	MP3, WMA9, AC-3, AAC, ATRAC3plus	MP3 y MPEG4 AAC LC	MPEG4, AMR, ATRAC3plus, G.726, WMA, LPCM
Video	MPEG2	MPEG1, MPEG4, WMV9	MPEG4 AVC (AAC LC Asoc. Audio)	VC1, H.263, MPEG4 part 2, MPEG2, MPEG4 AVC

1.5.3.1 Media Format

The media format describes the way to encode and the format for each one of the three kinds of media: audio, video and video with audio (AV). The term format is equivalent to codec or codec family.

The media format model is intended to achieve interoperability on the network, while the innovation in the media codec technology goes on. It defines a set of media formats and a set of optional media formats for audio, video and AV. DLNA also provides rules for the use of optional formats between compatible devices and converts optional formats into mandatory ones and vice versa. In the following table, we can see both the mandatory and optional formats for fixed and mobile household appliances (Table 1.6).

1.5.3.2 Media Transport

Media transport defines how the data move through the network. The grounding of the DLNA transport for any device that deals with media content through the network is HTTP 1.1. It is necessary to use this protocol, but there is also an optional protocol of transport in DLNA, namely the real-time transport protocol (RTP).

1.5.3.3 Management of Media, Distribution and Control

Media managing allows devices and applications to identify, manage and distribute digital content across devices on the network. UPnP technology AV is the solution for the management and control of devices developed according to the guidelines for the interoperability of devices on the network. UPnP AV architecture allows devices to support the entertainment content in any format and in any transport protocol. The services provided by this technology are:

- ***Content Directory Service***. This service provides a mechanism for each content server on the network as well as a standard directory and all its available content to any interested device. It enumerates the content and presents a logic structure for the multimedia library available on the server, such as videos, music and images.

- ***Connection Manager Service***. Determines the way the content can be transferred from the media server to a media player device. This service is used to carry out one of the following actions:
  - Match the capacity between the server and player devices.
  - Set up and remove connections between devices.
  - Find out information about current transfers on the network. When connections are made, the connection manager service is the interface between the devices and the TCP/IP stack.
- ***AV Transport***. This controls the flow of audio and video including the functions of play, stop, pause and search.
- ***Delivering Control Service***. Many devices contain attributes that can be configured dynamically. They make differences in content delivery, such as brightness and contrast in video devices or volume, balance and the equalizer in audio devices. This allows the control point to discover the attributes that support a device and retrieve, change and restore the configuration of any of these attributes.

Figure 1.34 shows the typical sequence when reproducing multimedia content.

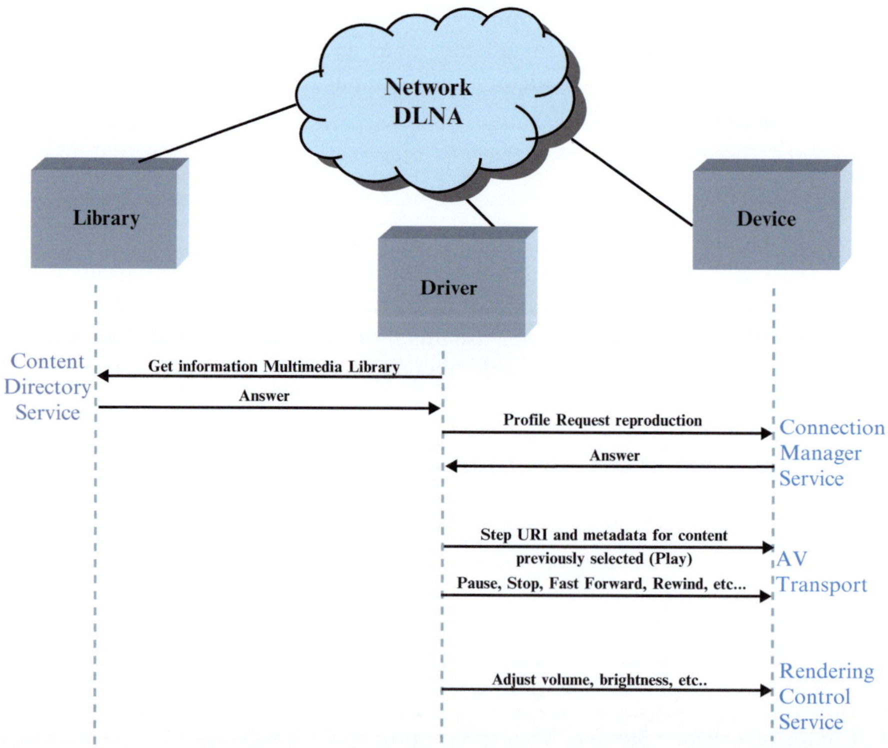

Fig. 1.34 Sequence of actions

1.5.4 *Penetration in the Market*

DLNA is currently implemented in the home in the usual way, especially with the appearance of lots of important manufacturers of devices that incorporate this technology. Among the most common devices using DLNA and incorporating it into our home are TV sets (with 400 certifications in the second quarter of 2009), games consoles, mobile phones (such as Nokia N95, which incorporates this standard), players and even cameras. There were 2,000 certified devices during the first half of 2009. Thus, it seems that this standard is becoming more and more relevant for the exchange of information and interactivity between terminals.

Attempts have been made to expand the DLNA domain further so a device will be able to connect to any network. For example, Oh Yeon-Joo et al. (2007) implemented the DLNA proxy server to service any virtual network (Ferguson and Huston 1998).

1.6 Other Standards

1.6.1 *Salutation*

This platform is independent of the architecture, language and operating system on which it is installed. It is based on the operation of the translation manager, specific for the Runtime Environment, and the salutation manager, which provides an API for publishing and search services (Suri et al. 2003). For example, Miller and Pascoe (1999) demonstrated the integration of this protocol with Bluetooth Service Discovery.

1.6.2 *Service Location Protocol*

This was created for client/server applications and it defines three kinds of agents: user, service and directory (Veizades et al. 1997). For more information, many of the protocols discussed in this chapter are compared and classified in Bettstetter and Renner (2000) and Zhu et al. (2002).

1.6.3 *Ad hoc Developments*

Before the appearances of concepts related to the automatic installation of devices in distributed networks, algorithms were developed for specific types of robots. This is the case for the Multi-Robot System of UNIX, which uses TCP/IP connections in a client/server architecture.

Standards have also been created for a particular type of technology such as the service discovery protocol (Avancha et al. 2001). This protocol can discover information on existing services in other Bluetooth devices.

Chapter 2
Robotic Development

Pablo Gómez del Torno, Omar Álvarez Fres, and Samuel Marcos Pablos

Abstract The complexity involved in the development of a project with several robots in a changing and unstructured environment often requires the use of many and various development tools. Not only complexity but also economic issues force the use of simulation tools to reduce costs. This chapter highlights some of the most relevant tools for robotic software development. These tools are mainly designed for the field of mobile robotics but some of them could be used for other kinds of robots. Some simulation tools belong to a specific robotic platform but the most powerful tools span multiple robotic platforms. The knowledge of the existence of these tools and their characteristics can make a big difference to the development time of a project.

2.1 Introduction

Robotic applications are constantly improving their complexity and functionality. With the advance of information technologies and engineering, robots are becoming more and more common tools in our workplaces and homes. This is the reason why it is necessary to develop a middleware that provides clear contexts, predefined data structures, blocks of code, standard communication protocols, synchronization mechanisms and so on (Cañas et al. 2006).

With the heterogeneous development of robots, different middleware platforms have emerged; in some cases, the manufacturers themselves have developed these

P.G. del Torno (✉) • O.Á. Fres
Infobotica Research Group, University of Oviedo, Oviedo, Spain
e-mail: gomeztpablo@uniovi.es; UO1475@uniovi.es; alvarezomar@uniovi.es

S.M. Pablos
Fundación Cartif, Valladolid, Spain
e-mail: sammar@cartif.es

I.G. Alonso et al., *Service Robotics within the Digital Home*, Intelligent Systems, Control and Automation: Science and Engineering 53, DOI 10.1007/978-94-007-1491-5_2,
© Springer Science+Business Media B.V. 2011

platforms to program their own products (Sony offers OPEN-R for Aibo, iRobot offers its Mobility Robot Integration Software (iRobot Corp. 2000) for its robots B12–B14, etc.). By contrast, research groups have developed platforms designed to cover their needs including:

- CARMEN (Carnegie Mellon Navigation Toolkit) (Montemerlo et al. 2003) developed by Carnegie Mellon University. OROCOS (Open RObot COntrol Software) (Bruyninckx 2003) developed by the Catholic University of Leuven. Player/Stage/Gazebo Project developed, at its beginnings, by the University of South Carolina (Gerkey et al. 2003).
- Miro (Middleware for Mobile Robot Applications) developed by the University of Ulm (Utz et al. 2002).
- MARIE (Mobile and Autonomous Robotics Integration Environment) developed by the University of Sherbrooke (Côté et al. 2004).
- Webots developed by the Swiss Institute of Technology in Lausanne and Cyberbotics Ltd. (Michel 2004).

In the following section, we will explain in more detail some robotic simulators. Generally these robotic platforms for programming are distributed as free software and are intended to be universal, i.e. platforms that support robots from any manufacturers.

In addition to studying the behavior of the robots, a thorough study of their communications and specifications (attenuation, power, wiring or radio frequency) is imperative. For this, some protocol simulators will be analyzed in the following sections. Protocol simulators are software tools that emulate communication networks and return data about network performance.

The aim of development platforms is to simplify the creation of robotic applications; the choice of a specific platform is often decisive for the proper integration of all elements to achieve a good level of performance, efficiency and reusability along with good communication between devices.

The reason why development platforms are used is that they allow the developer to obtain real data by making simulations of the behavior and communications of the robot in virtual environments. This has a direct impact on the reduction of costs since they shorten the times to develop all the activities and there is no need to have the robot, just a model of the simulator.

2.1.1 General Characteristics of Development Platforms

The wide variety of development platforms allows them to run under different operating systems (Windows, UNIX, Mac OSX, Linux) and to be implemented under several programming languages (C, C++, C #, Java, Python, LISP, Ada, Octave, Ruby, Scheme). Development platforms use different libraries; some of which are described below.

2.1.1.1 Standard Template Library (STL)

STL provides containers, iterators, algorithms and functions. STL (Stepanov and Lee 1995) provides a set of common classes in C++ (such as containers and associative arrays) that can be used with any compiler and supports some elementary operations (such as copying and assigning). STL algorithms are independent of the containers. This fact reduces the complexity of the libraries. The STL achieves its results using templates. This approach provides polymorphism in compiling time, which is more efficient than is the commonly used runtime polymorphism.

2.1.1.2 Microsoft Foundation Class Library (MFC)

MFC (Holzner 1993) is a library that contains, in a set of C++ classes, Windows APIs, thereby achieving easier access to them. The classes are defined by Windows Object handlers, predefined windows and common controls. The development of this library has been made in conjunction with new versions of a Visual C++ programming environment.

2.1.1.3 Open Graphics Library (OpenGL)

Silicon Graphics Inc. developed, in 1992, a standard specification defining a multi-language and multiplatform API to write applications that produce 2D and 3D graphics called OpenGL (OpenGL 2010). The interface consists of over 250 different functions that can be used to draw complex 3D scenes from simple geometric primitives such as points, lines and triangles. Its use extends to CAD applications, virtual reality, scientific visualization, information visualization and flight simulation. It is also used in game development, where it competes with Direct3D on Microsoft Windows platforms. In addition, there are several helper functions or class structures.

2.1.2 Robotic Middleware and Development Platforms

We will focus on the following middleware and development platforms.

2.1.2.1 CARMEN (Carnegie Mellon Navigation Toolkit)

This was developed in 2007 by Carnegie Mellon University (EEUU) (Carnegie Mellon University Home 2010) as a collection of robot control software in open source. CARMEN is designed to provide a consistent interface and a set of primitives for robotic application development in a wide variety of commercial robot platforms. The goals of CARMEN (Montemerlo et al. 2003) are to eliminate the barriers for the implementation of new algorithms for real and simulated robots and to facilitate the exchange of research and algorithms between different institutions. The purpose of this platform is not focused on adopting a strict standard, but on recommending good design methods to developers. CARMEN is a modular software architecture organized in three levels:

- The basic layer is responsible for the interaction and control of hardware; it provides an abstract configuration of the base and sensor interfaces. Likewise, it also provides a low-level control of movement in a straight line or simple rotations, a low-level collision detection and information from motion sensors with the aim of improving the operation of the odometers. The basic control modules of CARMEN can be implemented in a wide range of commercial robots such as Nomadic Technologies Scout and XR4000, ActivMedia Pioneers, iRobot b21 and the ATRV series.
- The navigation layer implements primitives including location, dynamic object tracking and motion planning. Unlike other navigation systems, CARMEN integrates in a single module all the motion control, except low-level motor control.
- The third layer is reserved for user-level tasks using primitives of the second layer.

2.1.2.2 Miro (Middleware for Mobile Robot Applications)

Ulm University (Germany) (Universität Ulm Home 2010) developed an article called Miro (Middleware for Mobile Robot Applications) (Utz et al. 2002) in 2002. In this article, they research the building of a robotic object-oriented middleware capable of making the development of applications for mobile robots easier and faster to promote the portability and maintainability of robot software. Miro has been designed and implemented to meet the requirements for the object-oriented standard of CORBA (OMG's CORBA 2010). The functional core of Miro and the data processing routines of sensors and control actuators are completely implemented in C++ allowing this way a high running efficiency. Miro is structured in a three-layer architecture related to the two main layers of CORBA:

- The device layer provides interface abstractions oriented to objects for all sensors and actuators of the robot. This is the part of Miro that depends on the platform used for hardware.
- The services layer provides a definition of the services available in sensors and actuators using an IDL (Interface Definition Language) from CORBA and implements these services and platform-independent objects.
- The Miro class framework provides a set of functional modules that are often used for mobile robot control, such as modules for mapping, localization, behavior generation, access path planning, registering and viewing facilities.

2.1.2.3 OROCOS (Open RObot COntrol Software)

The Catholic University of Leuven (Belgium) (Katholieke Universiteit Leuven Home 2010) published in 2001 an article in which it was shown the development of the project OROCOS (Bruyninckx 2003). OROCOS emerges as an open source platform with the following objectives:

- Open source license.
- High modularity and flexibility.
- The highest quality from a scientific point of view, based on its documentation and its technical structure.
- Independence from commercial robot manufacturers. Adapted to all robotic devices and computing platforms as well as multilanguage.

The OROCOS codebase is divided into modules or libraries. There are three main libraries:

- Support module. This software is without functional content for robots and includes 3D visualization and simulation, a tool for software configuration for components, a system of real time operation, communication between processes, documentation writing tools, and so on.
- Robotic module. This software implements specific algorithms for robots, kinematics and dynamics of servo-motors, serial and parallel manipulators, and so on. It makes use of one or more supporting modules.
- Components. These are CORBA objects, described using an IDL.

2.1.2.4 Player

Player is a network server for robot control (see Sect. 2.2). It provides a clean and simple interface to the robot's sensors and actuators over the IP network. The client program communicates with Player over a TCP socket, reading sensory data, writing commands to actuators and configuring devices on the fly.

2.1.2.5 Urbi

Urbi is an open source software platform used to control robots or complex systems in general. Urbi includes UObject, a C++ component library with a robot standard API that can match components to be used seamlessly in highly concurrent settings (Urbi 2010). The objective of Urbi is to help make robots compatible and simplify the process of developing software and behaviors for those robots.

Urbi simplifies the orchestration of independent concurrent components. It provides features to coordinate the execution of various components (e.g., actuators, sensors and software devices that provide features such as text-to-speech, face recognition and so forth). Languages such as C++ are well suited to program the local, low-level handling of these hardware or software devices; indeed, these need efficiency, a small memory footprint and access to low-level hardware details (Gostai 2010).

Urbi has an orchestration language to join different components and so describe high-level behaviors, namely urbiscript. Urbiscript is a programming language primarily designed for robotics. It is a dynamic, prototype-based, object-oriented scripting language that supports and emphasizes parallel and event-based programming by providing core primitives and language constructs. The urbiscript language syntax is very close to C++ syntax and is fully integrated with C++.

2.1.2.6 Orca

Orca is an open source middleware framework for developing component-based robotics. It is designed to target applications from single vehicles to distributed sensor networks. The main goals of Orca are to enable software reuse in robotics, simplify that software reuse and encourage that reuse of the software.

Orca enables the implementation of a distributed component-based robotic system by allowing the user to define interfaces and communication mechanisms. It was implemented using CORBA (Fumio et al. 2004), and it supports different programming languages such as Java, C# and C++.

2.1.2.7 OpenRDK

OpenRDK is a modular software framework focused on the rapid development of distributed robotic systems (also with heterogeneous robots) (OpenRDK 2010). In this framework, the main entity is a software process called an agent. A module is a single thread inside the agent process; modules can be loaded and started dynamically once the agent process is running (RoSta 2010a).

Modules communicate using a blackboard-type object (see Sect. 2.2.3.2), in which they publish some of their internal variables or properties. The access to remote properties is transparent from a module perspective. This also reduces shared memory (OpenRDK provides easy built-ins for concurrency management) in the case of local properties.

2.1.2.8 CLARAty

The CLARAty (Coupled Layer Architecture for Robotic Autonomy) is a robotic software framework to aid engineers to develop robotic applications.

CLARAty has a two-layer architecture that is designed to improve the modularity of its system software. This alternative is an evolution of the conventional three-layer architecture. The new architecture joins the planner and the executive levels of

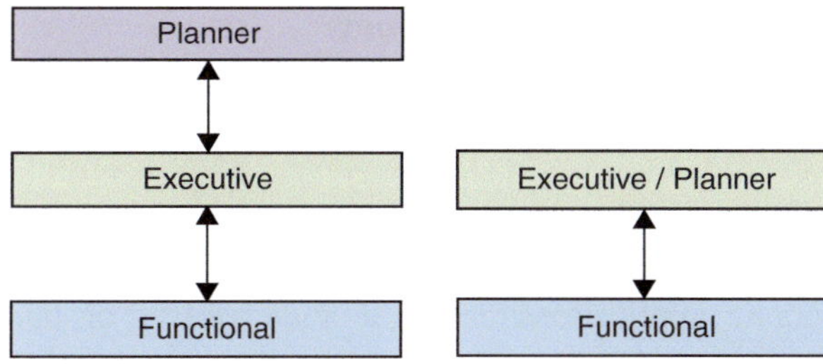

Fig. 2.1 Common three-layer architecture and proposed second-layer

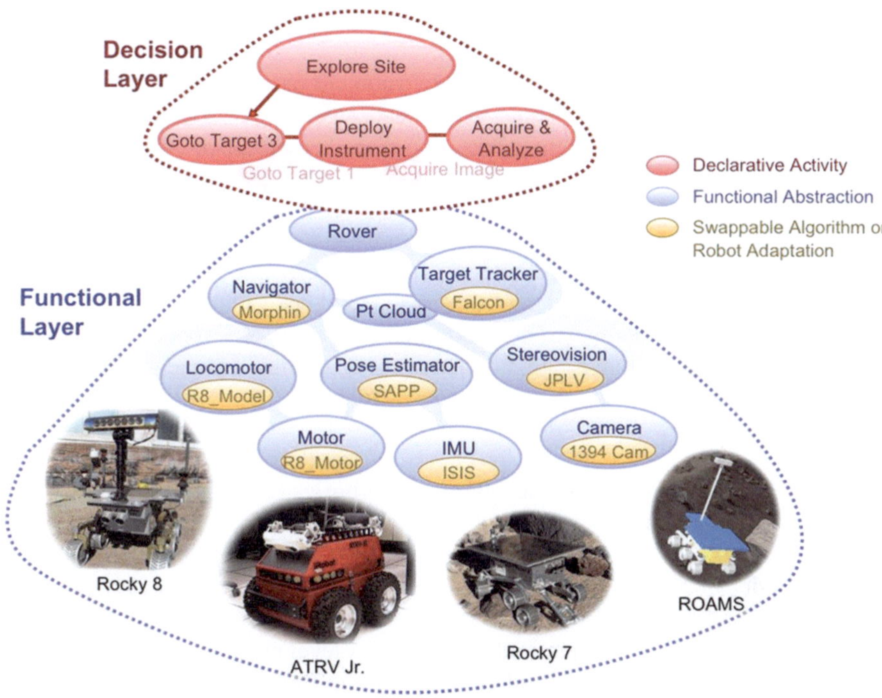

Fig. 2.2 CLARAty architecture with a functional layer and a decision layer

the conventional architecture in one decision layer (Volpe et al. 2001). One difference between the two architectures is the distinction between the levels of granularity and levels of intelligence (Fig. 2.1).

The functional layer is an interface for all system hardware and capabilities through which the decision layer uses the robotic system. The decision layer is an engine that is used to assess system resources and mission constraints. This layer includes planners, executives, schedulers, activity databases and planner-specific heuristics (Nesnas et al. 2003; Fig. 2.2).

2.1.3 Robotic Simulators

2.1.3.1 Microsoft Robotics Studio (MSRS)

The Microsoft Robotics Developer Studio (see Sect. 2.3) is a Windows-based environment for academic, hobbyist and commercial developers to easily create robotic applications across a wide variety of hardware (Microsoft Robotics Developer 2010).

2.1.3.2 Webots

Webots (see Sect. 2.5) is a development environment used to model, program and simulate mobile robots. With Webots, users can design complex robotic setups, with one or several similar or different robots, in a shared environment.

2.1.3.3 Stage/Gazebo

Stage (see Sect. 2.2) is a simulator commonly used with Player that simulates a population of mobile robots, sensors and objects in a 2D bitmapped environment. Gazebo (see Sect. 2.2), like Stage, is capable of simulating a population of robots, sensors and objects, but does so in a 3D world (Player 2010).

2.1.3.4 MARIE (Mobile and Autonomous Robotics Integration Environment)

The University of Sherbrooke (Canada) (Université de Sherbrooke 2010) considered in 2004 (Côté et al. 2004) a tool for the reuse of code for programming mobile robots. MARIE created an environment for system-level programming, simplifying the reuse of applications, tools and environments programmed in a coherent and integrated system.

2.1.3.5 AnyKode Marilou

AnyKode Marilou (AnyKode 2010) is software based on MSRS for modeling, programming and simulating an environment for mobile robots. The programming languages supported are C/C++, VB#, J#, C#, C++ and CLI and the programming can be under Windows or Linux.

The simulation of AnyKode Marilou supports two operating modes, real simulation or accelerated simulation, and supports multiple robots. The platform includes libraries for embedded robotic components: motors, servo-motors, odometers, force/torque sensors, distance sensors (US, IR, laser), laser range

Fig. 2.3 Scenario of the AnyKode Marilou software

finders, bumpers, air pressure forces, cameras, panoramic spherical cameras, GPS, accelerometers/gyroscopes and more (Fig. 2.3).

2.1.3.6 USARSim

This is a simulation system originally designed at the Carnegie Mellon University (CMU) and the University of Pittsburg. Initially, USARSim was focused on urban search and rescue, but has evolved into a general purpose simulation system (see Fig. 2.4).

USARSim was built on the engine (Unreal Engine) of the popular game Unreal Tournament, beginning with its first version. USARSim is a tool for early testing and late binding, and it can be used to verify the impact of the desired choices on a virtual environment and predict the behavior of the real system. The simulator includes several models of sensors (e.g., odometry, sonar, omnidirectional camera), robots (e.g., Kenaf robot, P2AT, Snow Storm) and actuators (Balaguer et al. 2008).

2.1.3.7 EyeSim/EyeBot

EyeSim is a 2D-specific simulator for the EyeBot mobile robot system. The simulator was not implemented as an independent program or process, and

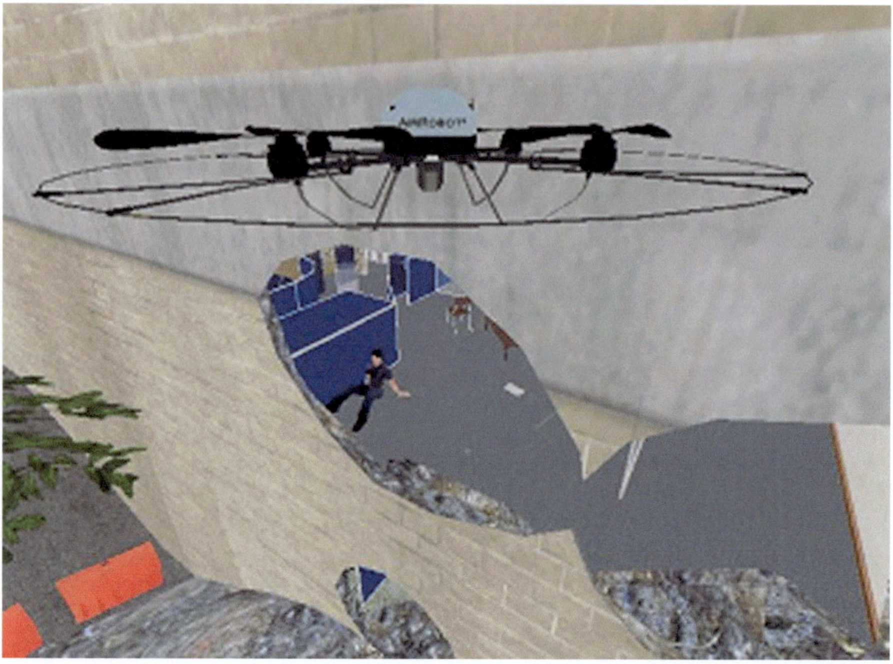

Fig. 2.4 Scenario of the USARSim with a robot (AirRobot) acting in a danger situation

differs from most existing simulation platforms because it was implemented as a library, which is linked to the robot application program (Bräunl and Graf 2008) (see Fig. 2.5).

The EyeBot is a controller for mobile robots with wheels, walking robots or flying robots. It consists of a powerful 32-bit microcontroller board with a graphics display and a digital grayscale or color camera.

2.1.3.8 MobileSim

MobileSim is software for simulating mobile robots and their environments and for debugging and experimenting with ARIA or other software that supports MobileRobots platforms (MobileRobots 2010).

MobileSim builds a Stage environment from a MobileRobots/ActivMedia and places a simulated robot model in that environment. It then provides a simulated Pioneer control connection via a TCP port. ARIA is able to connect to TCP ports instead of serial ports (ArSimpleConnector, for example, automatically tries TCP port 8101 before the serial port). MobileSim is based on the Stage library and has the GNU GPL license. Then, the most widely used robotic simulators will be studied more deeply: Player/Stage/Gazebo, Microsoft Robotics Developer Studio and Webots.

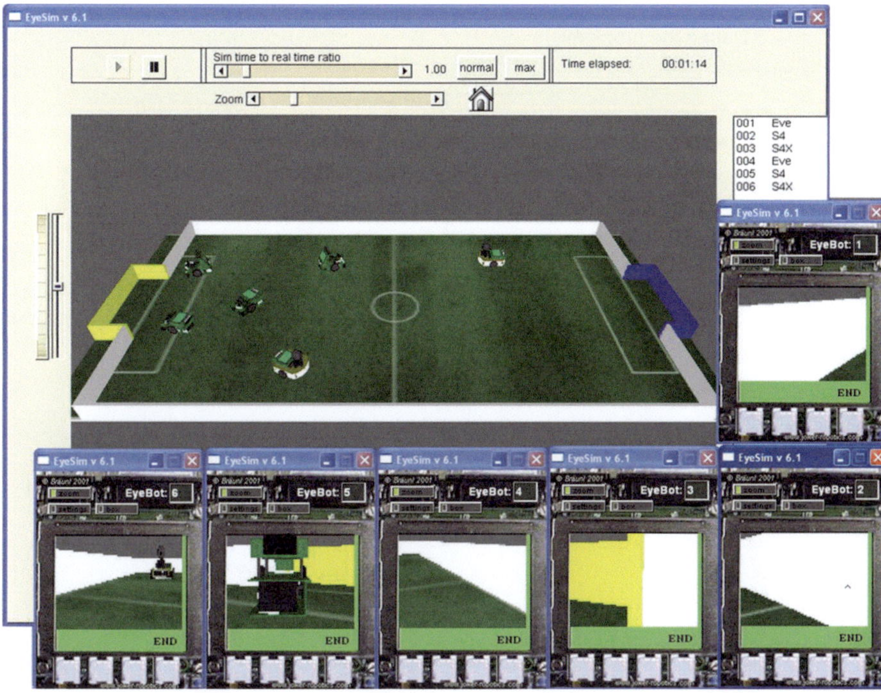

Fig. 2.5 EyeSim scenario

2.1.4 *Simulators for Communication Protocols*

2.1.4.1 OPNET Modeler

OPNET (Optimized Network Engineering Tool) is a development environment for the specification, simulation and performance analysis of communication networks. It can be simulated from small LANs to global satellite networks OPNET keys are (Chang 1999):

- Modeling and simulating. OPNET provides powerful tools to help users go through three of the five phases in a design cycle (see Fig. 2.6).
- Hierarchical model. OPNET employs a hierarchical structure for modeling. Each level of the hierarchy describes different aspects of the whole system being simulated.
- Specialized in communication networks. OPNET provides a detailed library of models for existing protocols and enables developers and researchers to modify any of these existing models or to develop new models.
- Automatic generation of simulations. OPNET models can be compiled into executable code. A discrete event simulation can be debugged or simply executed, resulting in output data.

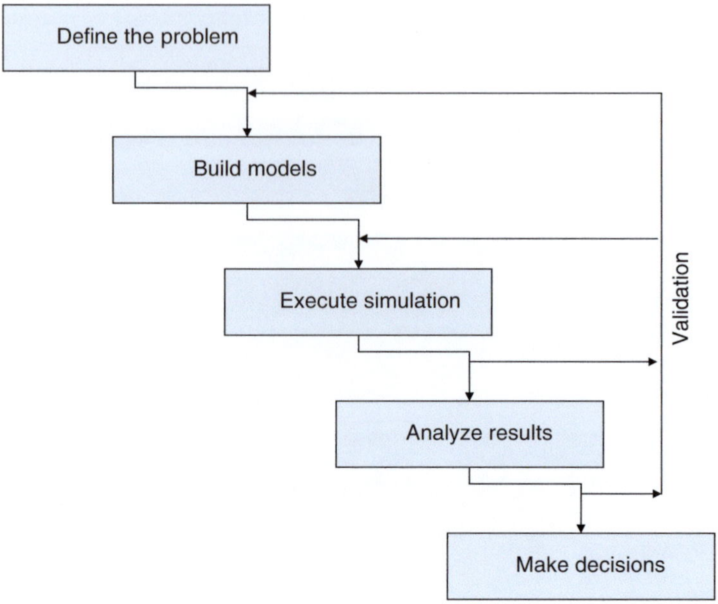

Fig. 2.6 Cycle of modeling and simulating

OPNET has four tools, called editors, to develop a representation of a system being modeled. Among these four tools, the parameter editor is considered a utilities publisher and not a publisher within the modeling domain. This is the reason why we focus on the three other editors (Brown and Christianson 2005):

- Network editor or project editor. The network editor is used to specify the physical network topology, i.e., it defines the position of the nodes and the interconnections through links. The specifications of each node are made in a lower layer. Nodes can be mobile, fixed or satellite peer-to-peer connections simplex or duplex (Fig. 2.7).
- Node editor. Communication devices created and interconnected at a network level need their specifications in the domain of the node. Node models are expressed as interconnected modules. These modules can be grouped into two categories: modules that have predefined characteristics and a set of construction parameters and programmable modules, which can be processors or queues specified in the processes editor.
- Processes editor. Models created by the publisher of processes are used to describe the logical flows between processors and queues. They are expressed in a programming language called Proto-C, which is a state transition diagram, a library Kernel and the standard language C.

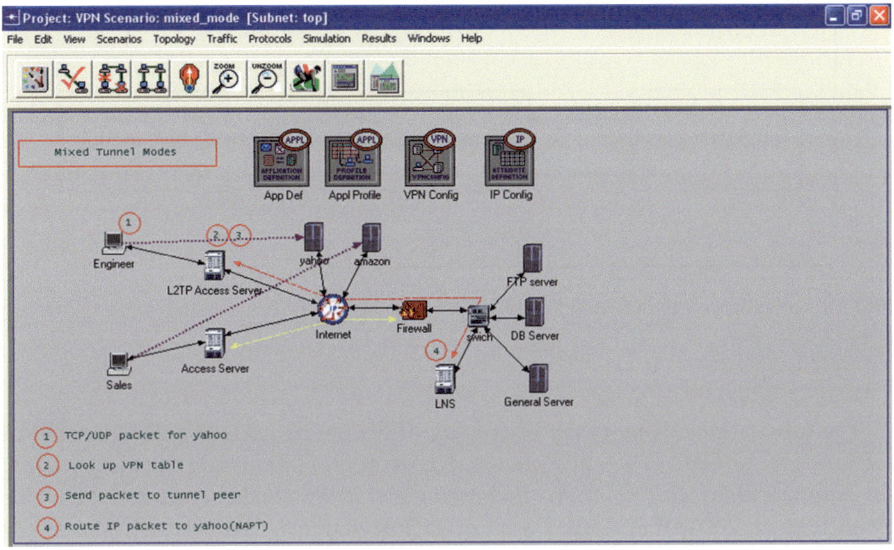

Fig. 2.7 Project editor of the OPNET modeler

2.1.4.2 OMNeT++

OMNeT++ is a discrete event simulator based on C++ for modeling communication networks, multiprocessors and other distributed or parallel systems. It was developed in 1992 by András Varga (Varga 2001), but came to light in September 1997. Simulcraft Inc. is in charge of selling the business license and providing support and consulting services on OMNeT++. OMNeT++ is an open platform and it can be used under GNU license.

The main application area of OMNeT++ is the simulation of communication networks, but because of its generic and flexible architecture, it is also successfully used in other areas such as the simulation of complex IT systems, queuing networks or hardware architectures. The motivations that led to its development were to obtain a powerful tool for discrete event simulation and an open platform for academic, educational or research-oriented use.

OMNeT++ tries to fill the gap between open source platforms for software simulation exclusively oriented to research studies, such as Network Simulator (Bajaj et al. 1999) and the alternatives in the market such as OPNET (OPNET 2010) that are too expensive.

OMNeT++ provides a component architecture for models. The components (modules) are programmed in C++, and then assembled into larger components and models that use a high-level language (NED). OMNeT++ can run on UNIX and Windows using Cygwin or the compiler of Microsoft Visual C++.

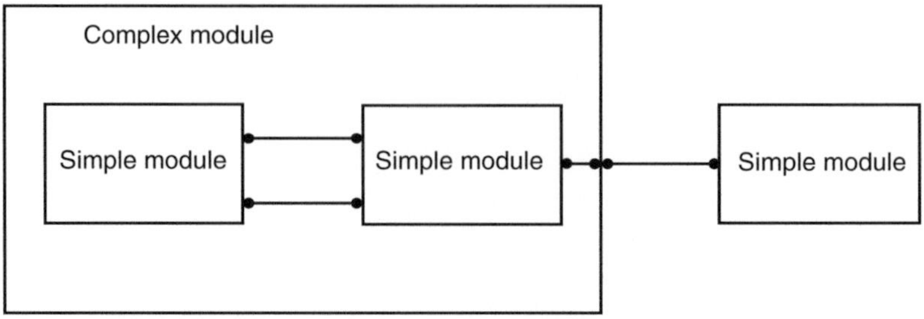

Fig. 2.8 Structure model in OMNeT++. The *arrows* represent the connections and the points are connecting doors

The use of OMNeT++ in the University of Karlsruhe led to the development of the projects *An OMNeT++ TCP model* (Kaage et al. 2001) and *A Simulation Suite for Internet Nodes with the Ability to Integrate Arbitrary Quality of Service Behavior* (Wehrle et al. 2001) that studied a set of TCP/IP models. Likewise, the University of Budapest carried out a project for the remote management of simulations in a workstation group using OMNeT++ (Erdei et al. 2001).

The design of OMNeT++ developed from the desire to support the simulation of large networks. This objective entailed some requirements (Varga and Hornig 2008):

- It must be able to perform large-scale simulations and hierarchical models and it has to be made of reusable components.
- The simulation software must emphasize simplifying the traceability and debugging in simulation models to reduce the debugging time.
- The simulation software must be modular, customizable and should allow the insertion of large simulation models in software applications such as network planning.
- The data interfaces need to be open and it must be possible to generate output and input files with the most popular software tools.

The OMNeT++ model is based on modules that communicate with each other through messages. The active modules, called simple modules, are written in C++ code and can be grouped into composed components and so on until we find the hierarchical level we want (Varga 2001). Messages can be sent through connections that link modules directly to the destination modules (see Fig. 2.8).

2.1.4.3 Network Simulator

Network Simulator (ns) is a tool used to simulate networks and was developed in 1995 (ns-1) by the network research group of the Lawrence Berkeley National Laboratory (LBNL's Network Research Group 2010) based on work with the REAL simulator (Keshav 1988). Network Simulator is capable of simulating several TCP types (including SACK, Tahoe and Reno) and queue algorithms in routers.

Although ns-1 based its simulation on Tcl (Tool Command Language) programs, its next version, 1996 (ns-2), used MIT Object Tcl and C++ language. The development of the simulator ns-2 was performed by DARPA VINT (Virtual InterNetwork Testbed) (VINT Project 1996) from 1997 to 2000; after this, it was developed by DARPA SAMAN (Simulation Augmented by Measurement and Analysis for Networks) (SAMAN 2001) and by NSF CONSER (Collaborative Simulation for Education and Research) (CONSER 2002) until 2004. Currently, its development relies on collaborators and volunteers and the project overall relies on Sourceforge.

The core of ns-2 is written in C++ but the simulation scripts are written in an extension of the object-oriented language Tcl. This structure allows simulations where a modification does not involve recompiling the simulator every time there is a structural change.

The ns-2 simulator has a tool to animate objects known as Network Animator and this is used to display the output of the simulator or to set up simulation scenarios graphically.

The next version of Network Simulator, ns-3, was developed by a group of researchers from the University of Washington, the Georgia Institute of Technology and the ICSI Center for Internet Research (The ICSI Networking Group 2010). Using the simulator ns-3, they built a discrete event simulator for the Internet network with educational and scientific purposes, with an emphasis on layers 2–4 of the network stack. It also has the following objectives (Henderson et al. 2006):

- The ns-3 project must adopt the methodology of a community oriented to open source.
- The simulator will be distributed freely as open source software and must be compatible with other open source networking software.
- The simulator must have a scalable, extensible and modular architecture, a clear design, good documentation and be capable of making emulations.
- The core of the models should be well tested and validated.
- The project should develop a series of simulation experiments that will be a canon for its current use on networks.

The simulator ns-3 can simulate the IPv4 and IPv6 networks as well as realistic models on different abstraction levels. It can be installed on Linux, OSX (Darwin), Windows (through emulation) and FreeBSD. Table 2.1 shows a specification of the network models existing in ns-2 and the additional ones of ns-3.

2.1.4.4 GloMoSim

GloMoSim (Global Mobile Information System Simulator) is a simulation environment for wireless networks and mobile devices (Zeng et al. 1998). It was developed by the UCLA Parallel Computing Laboratory (Bagrodia et al. 1998) between 1997 and 1999 and was made using PARSEC (Parallel Simulation Environment for Complex Systems), which was also developed by the UCLA Parallel Computing Laboratory and is based on the C language for parallel simulations.

Table 2.1 Differences between ns-2 and ns-3

Layers	ns-2	ns-3 (added to ns-2)
Aplication	Ping, vat, telnet, FTP, multicast FTP, HTTP, generation of probabilistic and traced traffic, webcache	Sockets as API (to port existing applications to the ns environment), peer-to-peer
Transport	TCP (several variants), UDP, SCTP, XCP, TFRC, RAR, RTP Multicast: PGM, SRM, RLM, PLM	Emulation of the TCP stack (Linux, BSD), DCCP, TCP variants with a different speed
Network	Unicast: IP, Mobile IP, distance vector and state of link, IPinIP, source routing, Nixvector Multicast: SRM, centralized MANET: AODV, DSR, DSDV, TORA, IMEP	Complete Support to IPv4, complete support to IPv6, NAT XORP/support to Click Routing: BGP, OSPF, RIP, IS-IS, PIM-SM, IGMP/MLD
Link	ARP, HDLC, GAF, MPLS, LDP, Diffserv Queues: DropTail, RED, RIO, WFQ, SRR, Semantic queue of packets, REM, Priority, VQ MACs: CSMA, 802.11b, 802.15.4 (WPAN), Aloha Satellite	New model 802.11, variants of 802.11 (mesh, QoS), 802.16 (WiMax), TDMA, CDMA, GPRS
Physic	Satellite Transponder, Power Model, omnidirectional antennas	Physical layers IEEE 802, GSM

GloMoSim is structured in layers in the same way as the seven-layer architecture from the OSI model, with standard APIs between each layer (see Fig. 2.9). This simplifies the implementation of new protocols and models at different layers (Farooq and Bilal 2006).

This design, based on layers, benefits modularity, i.e., the developer can implement new protocols on different layers without modifying the other layers. In addition, GloMoSim supports the parallel and sequential execution of discrete event simulation. Table 2.2 shows the protocols supported in the different layers of GloMoSim (Farooq and Bilal 2006).

2.1.4.5 IPC

IPC was developed in 1994 for the NASA New Millennium Program (NASA New Millennium Program 2010) and it has since been used in numerous robotic and autonomous systems at CMU, NASA and elsewhere. It is based on a previous CMU project called Task Control Architecture, which has also been used for NASA projects.

IPC provides high-level support for connecting processes using TCP/IP sockets and sending data between processes. It takes care of opening sockets, registering messages and sending and receiving messages, including both anonymous publish/subscribe- and client/server-type messages. The IPC library contains functions to

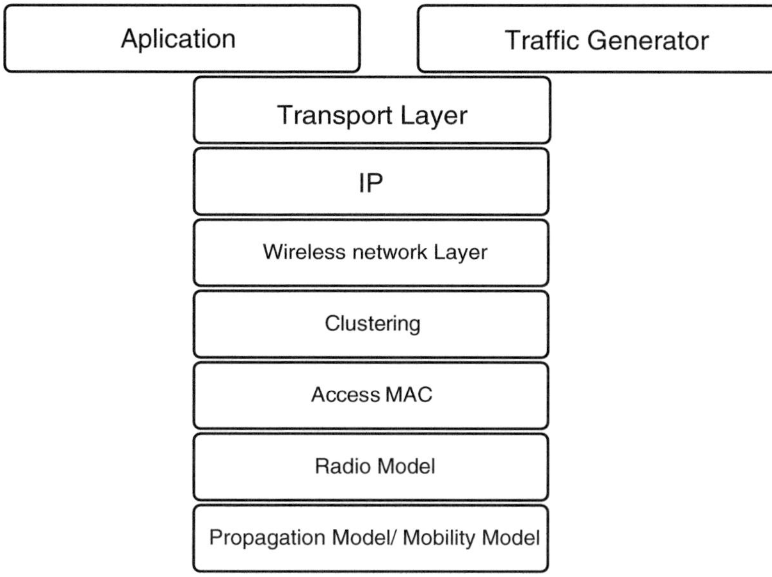

Fig. 2.9 Layer architecture of GloMoSim

Table 2.2 Protocols and models supported by every layer of GloMoSim

Layers	Protocols
Mobility	Random waypoint, random drunken, trace based
Radio propagation	Free-space, duplicated wave
Radio models	Cumulative noise
Models for reception of packets	SNR balanced, BER with BPSK/QPSK modulation
MAC	CSMA, IEEE 802.11 and MACA
Network (routing)	IP with AODV, Bellman-Ford, DSR, Fisheye, LAR scheme 1, ODMRP, WRP
Transport	TCP and UDP
Application	CBR, FTP, HTTP and Telnet

marshal (serialize) and unmarshal (deserialize) data, handles data transfer between machines with different Endian conventions, invokes user-defined handlers when a message is received and invokes user-defined callbacks at set intervals.

IPC libraries exist for C, C++ and Allegro Common Lisp and Java (tested only under Linux) (Allegro 2006). IPC currently runs on the following architectures and operating systems: Sparc (running SunOS and Solaris), Intel processors (running Linux, Windows NT, Windows 98), 680xx processors (running VxWorks), Silicon Graphics Inc. (running IRIX) and Macintosh (running Mac OS and OSX). It is easily ported to any machine that supports UNIX-style sockets (for assistance on porting IPC to a new architecture).

2.1.5 Numerical Simulation

Numerical or discrete simulation consists of reproducing simulations, typically using computers, over different processes to try to get a result as close as possible to the reality of the behavior of the signal or simulated process and thereby prevent potential problems and improve designs.

Numerical simulation is a useful procedure in RandD projects including structural, dynamic studies and the design of electrical circuits and fluid mechanics among others.

Specialized programs exist in each topic, but many processes can be simulated for any field. The only condition is to know their mathematical equations. The better these equations are, the closer the simulation is to reality. The most important programs in this field are Matlab (Simulink) and others for specific industries such as Cosmos (structural design), Fluent (Fluid Mechanics) and several specific branches including Abaqus (structural analysis and fluid mechanics).

The advantages include the possibility of conducting highly complex studies without having to make real scale models. In addition, a numerical simulation can be optimized progressively by adding more and more complex variables to achieve the expected results. All this can translate into huge cost savings.

The disadvantages highlight the importance of obtaining mathematical equations to describe as accurately as possible the process to be simulated. Otherwise, results might not reliably describe a simulated process. Another disadvantage is that it is necessary to choose the boundary conditions and an appropriate mesh to achieve its convergence.

2.1.6 Discussion

After analyzing all these development platforms, we will take two platforms that run on different operating systems. On one side, we will execute the robotic simulator Player/Stage/Gazebo on a Linux machine to perform simulations in virtual environments about the behavior of the robotic models under study. On the other side, we will simulate communications among robots, control points and several communication devices involved in our scenario using the protocol simulator ns-3. We have chosen Player/Stage/Gazebo because of its wide use in the world of robotics research, the availability of models to consider and its free distribution. The use of the ns-3 simulator is because of the multitude of models of protocols that can simulate and the free distribution of its GNU.

On another machine, we will install Windows and perform a robotic simulation using Microsoft Robotics Developer Studio because of its better graphics features to create virtual environments and its wide range of robotic models. Moreover, we will emulate the protocol simulator OMNeT++ under a Windows 32-bit platform to obtain different samples of the patterns of communication between robots and control devices.

We discarded the use of other robotic simulators because they did not fit our needs. We also discarded the use of other protocol simulators because of their cost (OPNET Modeler) or because they only focused on the field of wireless communications (GloMoSim).

2.2 Architectural Patterns for Robotic Development

Software development brings common problems that can be solved by different patterns. The patterns that define the structure of a software system are called the architectural patterns. It is useful to know and use these patterns in the development of software for robots to improve quality, maintainability and reusability as well as save time and effort (RoSta 2010a).

2.2.1 Layered View

This view is focused on decomposing a complex system into simpler parts.

2.2.1.1 Layers

This pattern is used to decouple components to support modifiability, portability and reusability. All components inside the same layer work at the same level of abstraction. Layers communicate with their adjacent layers through interfaces. A layer must communicate only with its adjacent layers. It is useful to decouple high-level from low-level responsibilities (RoSta 2010a).

2.2.1.2 Indirection Layer

The indirection layer is a layer situated between the interfaces and the instructions of a system. The aim of this function is to hide the subsystem providing services from the external world. This layer wraps all the relevant accesses to the system and performs other tasks such as converting and tracing invocations. This pattern can be implemented either as a part of the subsystem (as in a virtual machine) or as an independent entity (as in an adapter or facade pattern) (RoSta 2010a) (Fig. 2.10).

2.2.2 Data Flow View

This view is focused on processing and/or transforming data streams.

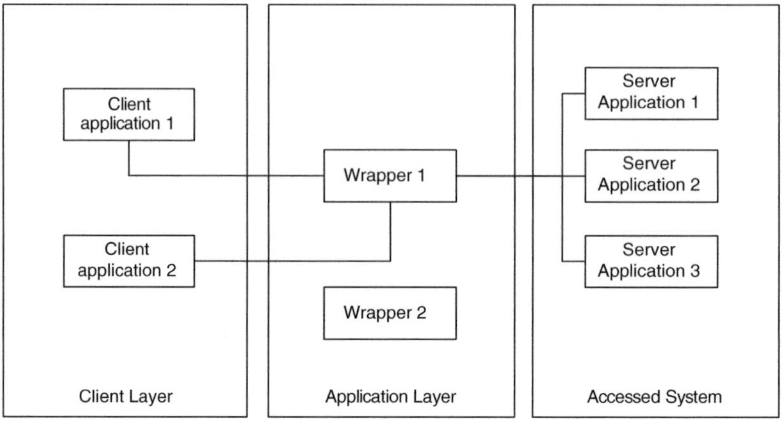

Fig. 2.10 Layer patterns

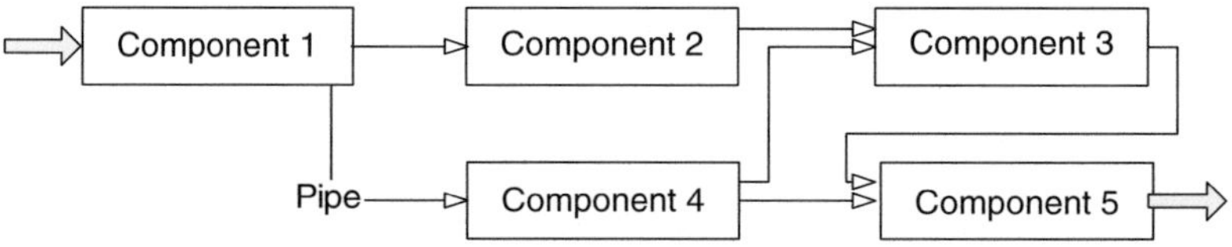

Fig. 2.11 Pipes and filters pattern

2.2.2.1 Batch Sequential

This is based on a divide and conquer algorithm, in which a complex task is divided into more simple sequential tasks or steps realized as independent components (filters). At each step, the data are processed and forwarded to the next component until completion (RoSta 2010a).

2.2.2.2 Pipes and Filters

As well as filters, data flow in several ways (pipes). As mentioned, at each step, the data are processed and forwarded to the next component until the completion (RoSta 2010a) (Fig. 2.11).

2.2.3 Data-Centered View

This view is focused on sharing information with a central repository of data that is accessed by multiple components.

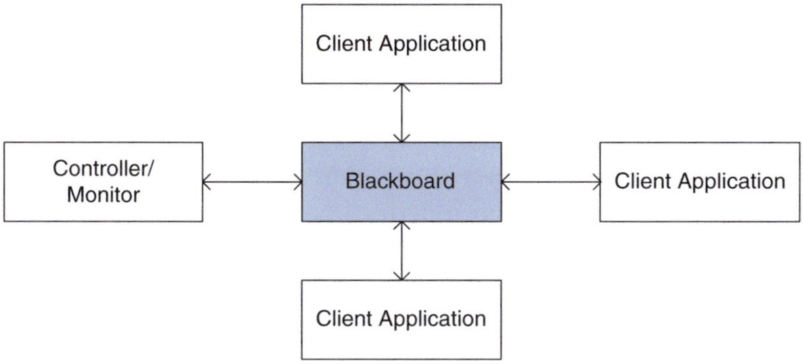

Fig. 2.12 Blackboard pattern

2.2.3.1 Active Repository

In this case, a shared repository (a central data store shared among other components) informs clients of any events occurring in the repository. This repository should be aware of the active clients to inform each client; communication often follows an event-based model, thereby reducing overheads (RoSta 2010a).

2.2.3.2 Blackboard

This solution proposes dividing a complex problem into smaller subtasks. Each subtask shares its computational data with the repository component. The repository component obtains the data from all clients to perform the necessary computations to improve the problem solution. A control component usually coordinates the clients according to the state of the repository component. This pattern is useful when no deterministic solution exists (RoSta 2010a) (Fig. 2.12).

2.2.4 Adaption View

This view is focused on the system adapting through its lifecycle/evolution. In this architectural pattern, the system is composed of two main parts: the core invariable component and the adaptable component(s).

2.2.4.1 Microkernel

The microkernel performs common/routine services that systems need to do. Clients can only access microkernel services through external server APIs. The microkernels are often structured in layers (RoSta 2010a) (Fig. 2.13).

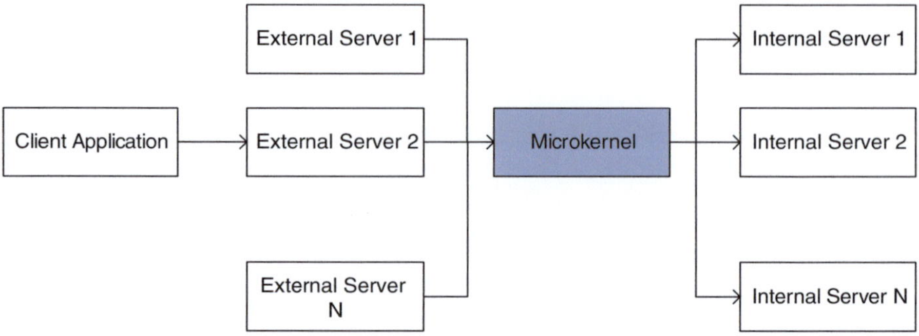

Fig. 2.13 Microkernel pattern

2.2.4.2 Reflection

In this architectural style, structural and behavioral aspects of a system are stored into meta-objects; these meta-objects (meta-object level) are separated from the application logic components (application logic level). This provides a mechanism for changing the structure and behavior of software systems dynamically (RoSta 2010a).

2.2.4.3 Interceptor

The interceptor's pattern is used to accommodate the system to future changes and to transparently update the services on runtime, thus increasing the system's maintainability. Therefore, this can ease adding or modifying an intermediate stage in the process.

2.2.5 Language Extensions View

This view is focused on disseminating components over a distributed environment to decouple components.

2.2.5.1 Interpreter

The interpreter is used when conversion is required at runtime. Requests are often parsed and executed within the same environment, namely the interpreter application. This provides good portability across many different systems. Common programming languages using interpretation include Python, Tcl, Perl and BASIC (RoSta 2010a).

2.2.5.2 Virtual Machine

A virtual machine performs similar tasks to the real hardware, but by using software. It executes an intermediate code, bytecode, rather than a machine code. A program written in a particular language is compiled into bytecode using a bytecode compiler. This architectural pattern offers portability across platforms (if virtual machine implementation exists). Examples of such systems include JVM, Parrot/Perl VM, Python VM and Microsoft VM (RoSta 2010a).

2.2.5.3 Rule-Based System

Rule-based systems consist of facts, rules and an engine that acts on them. Rules are applied to facts, which may lead to new facts and so on. Examples of such systems include Prolog language family systems or JBoss Drools Expert (RoSta 2010a).

2.2.6 Distribution View

This view is focused on disseminating components over a distributed environment to decouple components.

2.2.6.1 Broker

Broker architectural patterns can be used to structure distributed software systems with decoupled components that interact using remote service invocations. The broker is responsible for coordinating communication, addressing the requirements, transmitting results and exceptions. It is composed of several components responsible for request handling, invocation, requesting, marshaling the requests, client/proxy, and so on (RoSta 2010a).

2.3 Player/Stage

The Player/Stage platform was created by the researchers Brian Gerkey, Richard Vaughan, Andrew Howard and Nathan Koenig, who were members of the University of Southern California (Gerkey et al. 2003) during the 1990s; in 2001, it was moved to Sourceforge and emerged as a free development platform for the research of robotics and sensor systems.

Its development has since been expanded and the group of developers has been attached to working groups from different universities and institutions (Kranz et al. 2006). Nowadays, the similarity and versatility between models (real and virtual) place Player/Stage as a standard in open source communities for robotic research (Collett et al. 2005), with the support of more than 20 laboratories around the world (Vaughan et al. 2003). Its open platform nature makes it a tool in which users can collaborate; such is their acceptance that more than 200 users are part of the Sourceforge community (Player 2010). These include the Intelligent Cooperative Systems Laboratory from the University of Tokyo (Intelligent Cooperative Systems 2010), the Intelligent Autonomous Systems Group from the University of Munich (Intelligent Autonomous Systems Group 2010) and the Intel Corporation.

2.3.1 General Features

The Player/Stage Project is open source and widely used for robot control. It is divided into three parts (Player 2010):

- Player: This is a server for controlling devices and repositories of robots (sensors and actuators), divided into several libraries for more flexibility (Mohammed and Al-Jaroodi 2008). The client communicates with Player using a TCP socket that can access data, send commands or request changes to the configuration of a device in the repository.
- Stage: This is a multi-robot and bidimensional simulator capable of simulating a population of mobile robots, sensors and objects in an environment mapped in two dimensions. Stage is designed to support the study of multi-robot autonomous systems, so it provides simplicity in its structure. Simple computer models are simulated for each device instead of emulating the device with great fidelity. It pursues a temporary reduction approach for the approximate study of these multi-robot systems.
- Gazebo: This is a multi-robot simulator for 3D environments in open spaces. Like Stage, it is capable of simulating a population of robots, sensors and objects, but it does so in a world of three dimensions. It is capable of generating realistic and physically plausible feedback from the sensors and how they interact with objects, including an accurate simulation of rigid body physics.

Player/Stage/Gazebo is implemented in an architecture of three levels (Mohammed and Al-Jaroodi 2008):

1. The first level is made up of customers who are specifically developed as robot applications.
2. The second level is where Player provides common interfaces for various robot devices and services.
3. The third level is the current robots, sensors and actuators.

It is designed to be executed on Linux, Solaris, Mac OSX and Berkeley Software Distribution of Linux.

2.3.2 *Specific Features*

There are several specific features for each one of the tools in which Player/Stage is divided.

2.3.2.1 Player

Player is designed to be independent of the programming language and platform used. The client program can run on any machine networked to the robot and it can also be written in any language that supports TCP sockets. Any client can connect and read sensor data from any Player device on any robot (Player 2010). Player serves as an interface for many different types of robotic devices and provides drivers for many pieces of hardware. Each device in Player (Player 2010) is composed of:

- Drivers, which can be (Kranz et al. 2006):
  - A code that connects and communicates with a physical device.
  - An algorithm that receives data from another device and processes and returns them through the same channel.
  - A "virtual driver," which can arbitrarily create data when necessary.
- Interfaces, which are used by the client to write new applications that receive information from sensors or actuators.

Each interface is well defined; therefore, drivers only need to package the information into the appropriate interface format and send it to the client. Owing to the standardization of interfaces and the fact that Player/Stage is designed to be independent of language and platform, there are several utilities for the client, written in a wide variety of programming languages (C, C++, Java, Python, LISP, Ada, Octave, Ruby, Scheme) (Kranz et al. 2006).

2.3.2.2 Stage

Stage is a graphic simulator for mobile robots in two dimensions, which initially has two purposes (Gerkey et al. 2003):

1. To allow for the fast development of controllers for real robots.
2. To allow experimental robots to face possible scenarios for action.

Stage has been specifically designed to develop multi-robot systems; this feature allows us to experiment with a large number of robots in a virtual way, which means important savings compared with the real way of simulating with real robots (see Fig. 2.14).

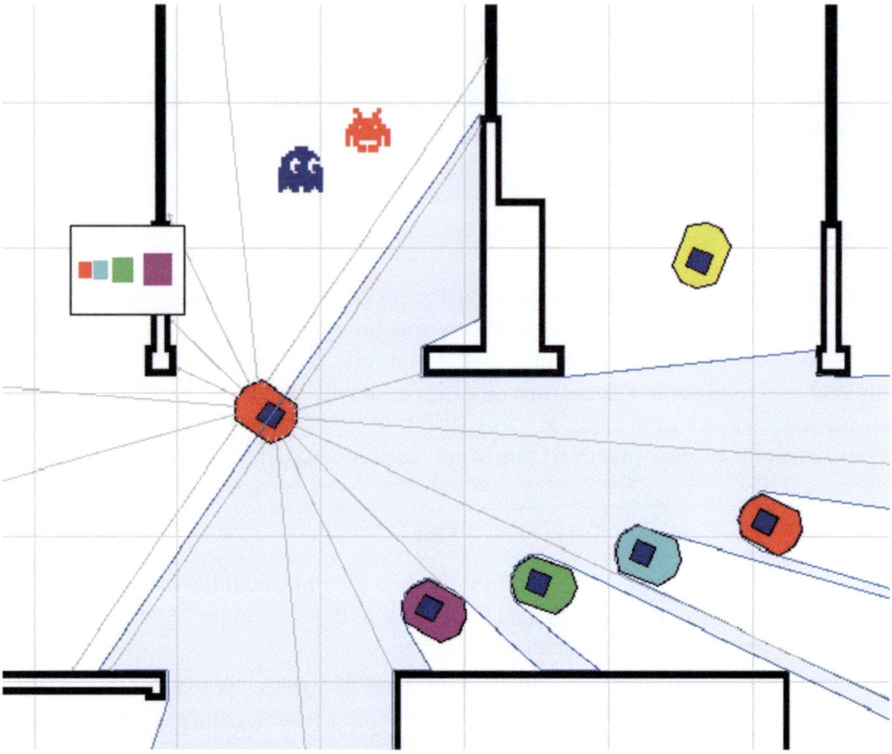

Fig. 2.14 Multi-robot simulation in stage (Player 2010)

The aspects that allow Stage to support multi-robot systems are:

- Fidelity: Stage provides package models of devices more cheaply than if users emulate each device with great fidelity.
- Lineal scaling with population: The algorithms used by the sensor models are independent of the number of simulated robots.
- Configurable: The device models are configurable and there is a wide variety of sensors, actuators, sonars and so forth.
- Player interface: There are hardly any differences between the simulated robots and real robots because of the use of standard interfaces.

There are two libraries in Stage:

- Libstageplugin. Stage is normally used as a plugin for Player. Users program controllers for robots and sensors and algorithms for clients for the Player server. This library provides a multitude of sensors and actuators, including laser and sonar ones, as well as a differential for the wheels of the robot.
- Libstage. Stage can also be used as a C library to provide a simulation of a robot in user programs. This utility is necessary if Player does not fit the needs of the developer or if we want to use a custom simulation model, based on a previously known simulation engine.

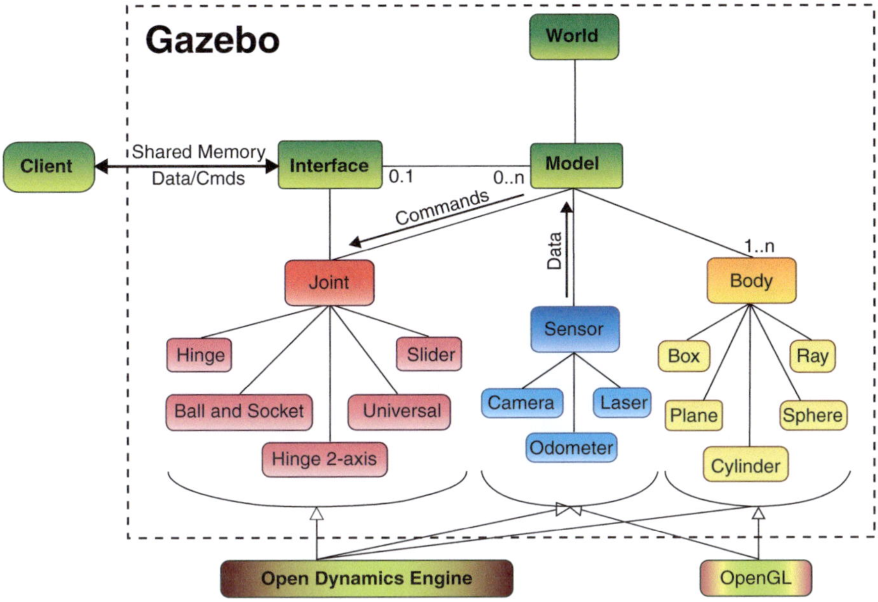

Fig. 2.15 Gazebo's software architecture (Hidalgo Bláquez and Cañas 2008)

2.3.2.3 Gazebo

Developed to be fully compatible with Player server devices, the Gazebo simulator is meant to be the 3D multi-robot platform for Player/Stage. It is capable of simulating a population of robots, sensors and objects in a virtual world in 3D. The Gazebo-simulated hardware is designed to reflect the behavior of its equivalent in reality (Koenig and Howard 2004); because of this, the client software uses an interface that looks identical to the real robot. This simulator is capable of generating physical situations such as the interaction among objects (including an accurate simulation of rigid body physics) (Player 2010). Gazebo's architecture (see Fig. 2.15) is based on the simulator engine ODE (Open Dynamics Engine) created by Russell Smith and belongs to the open source community (Open Dynamics Engine – home 2010). Gazebo uses OpenGL (OpenGL 2010) to render the images from simulated cameras or simply as a visualization tool (Beck et al. 2007). GLUT (OpenGL Utility Toolkit) is a tool based on OpenGL windows used in Gazebo for the visualization of simulations. The reasons for adopting this tool in Gazebo are its ease of use, light computational load and platform independence (Koenig and Howard 2004).

The main advantages of Gazebo are:

- The simulation of different position sensors such as sonar, laser scanning and GPS.
- It uses models of commonly used robots such as Pioneer2DX, Pioneer2AT and SegwayRMP.

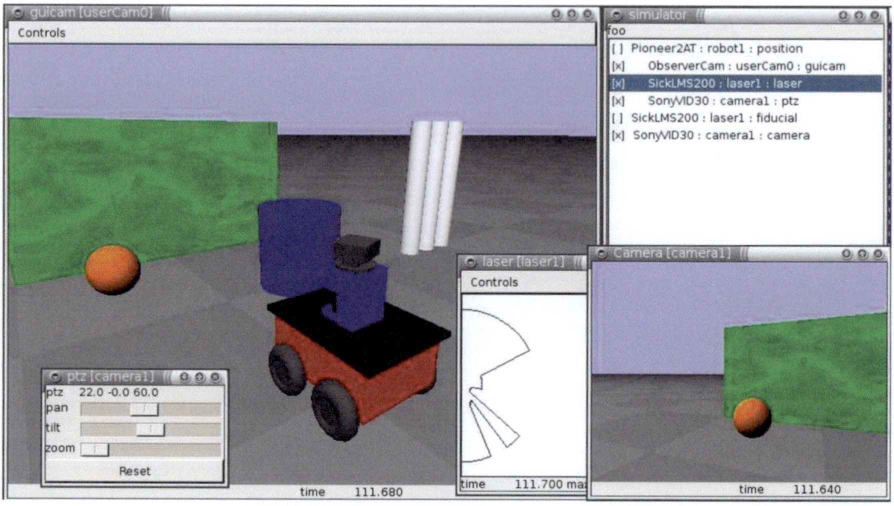

Fig. 2.16 Gazebo's graphic user interface (Player 2010)

- The realistic simulation of rigid body physics; robots can interact with objects (grab, push, etc.).
- Support for Player: Robots and sensors can be controlled through a standard Player interface.
- Independent operations: external programs to Player/Stage can interact directly with the simulator using the library libgazebo without having to go through Player.
- Graphical user interface written on wxPython (see Fig. 2.16).

2.3.2.4 Differences Between Stage and Gazebo

The Player/Stage simulator provides two multi-robot simulators; Stage and Gazebo are both compatible with Player. Client programs written using a simulator can be implemented in the other with only minor modifications. The key is in the philosophy of these two simulators' designs. Stage focuses on simulating large populations of robots with low fidelity, whereas Gazebo was designed to simulate small populations with high fidelity. It is for this reason that the two simulators can be considered complementary and they are used according to the needs of the researcher (Player 2010).

2.4 Microsoft Robotics Developer Studio

The Microsoft Initiative: Microsoft Robotics (Jackson 2007), created in December 2006, aims to become an industry standard for robot control, so it must overcome the many technical differences between the robots. It is said that a robot is a system that connects sensors and actuators via an electronic mean of communication. To interact with these robotic systems, MSRS provides the mapping between decoupled software modules and hardware components or subsystems of the robot (Chrysanthakopoulos and Nielsen 2007).

To be a friendly application to the user, MSRS also has a graphical interface where you can run different scenarios. It also has a lightweight runtime environment and includes a service-oriented routine. As shown below, it is an end-to-end development platform that allows programmers to create services for a great variety of robotic hardware.

2.4.1 General Features

The architecture design of MSRS follows the pattern for transferring states (Fielding 2000) and interacts with robots using software services (such as web services). These services are decoupled, which allows the reuse of the code. The interaction of services, particularly the control system, is defined through the use of a configuration file based on XML. This manifest file, written in XML, makes use of the functionality provided to identify each service available in MSRS. Each service can be expressed as a state machine and remain available for review through the network.

By contrast, the use of the SOAP (W3C 2007) allows the availability of a URL in which the state data can be viewed through any Internet browser. MSRS is implemented using .NET, so a .NET language is required to write directly into MSRS services.

Typically, the service implementations are written in C#, although any other language available in Visual Basic, C++ editors or Iron Python can also be used. In addition, through the use of SOAP interfaces, every department can communicate with other interfaces from different programming platforms with the drawbacks of requiring a more elaborate development and suffering losses in the encoding/decoding of messages.

There are two important limitations in this kind of implementation (Jackson 2007). First, MSRS is the ideal platform for managing real-time systems (RTS) since it requires a high frequency of monitoring. The problem comes from the fact that the system memory will be managed by .NET and this offers no guarantee of stability, so a particular service could be interrupted during periods of milliseconds.

The solution to this problem is keeping the RTS code running separately from the robot´s simulation and adding a gateway consisting of an MSRS service to communicate between the RTS and the robot. The second limitation is that MSRS requires that all services must be implemented in an environment where the entire .NET library is present, and some robots do not have processors capable of handling these environments.

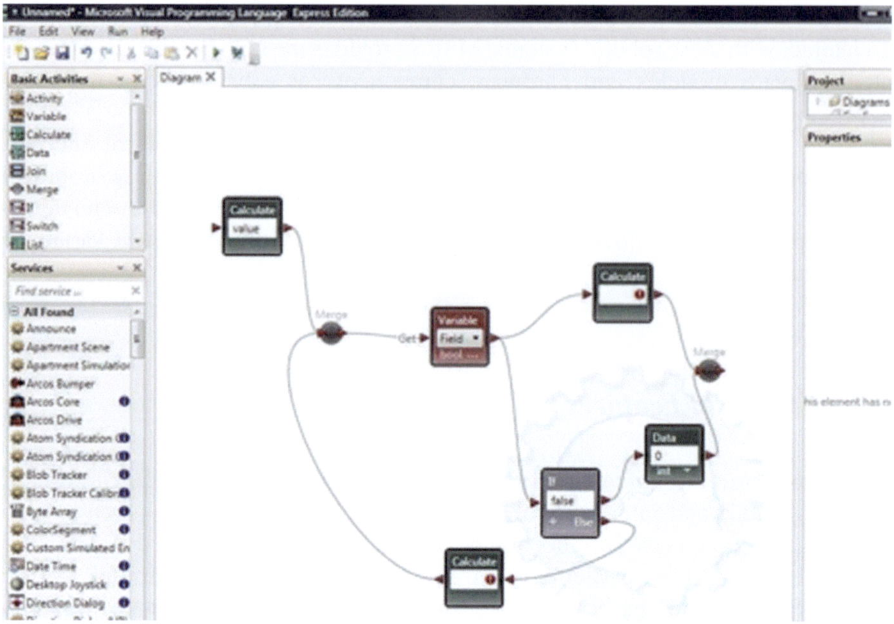

Fig. 2.17 VPL user's interface

2.4.2 *Specific Features*

The fact that MSRS is designed to create robotic applications for a variety of hardware platforms gives it special characteristics, such as the following (Microsoft Robotics 2010).

2.4.2.1 End-to-End Development Platform

MSRS allows developers to build services for a wide variety of robotic hardware components:

- Visual programming. MSRS has a visual programming language (VPL) that allows developers to create and test robotic applications intuitively. VPL is a programming environment based on graphic data flows instead of flow control. The programming data flow VPL consists of a sequence of activities represented as blocks (see Fig. 2.17) tied together. VPL can place several blocks together into one to work with it in our program.
- Simulation of robotic applications in 3D virtual environments. Microsoft Visual Simulation Environment (VSE) is a tool that includes MSRS to simulate the environments and robots in three dimensions (see Fig. 2.18). The development of robotics involves an effort in simulations to enable the test of robotic applications when they face a particular application environment. VSE includes AGEIA™ PhysX™,

Fig. 2.18 Scenario in VSE

which was firstly developed by AGEIA Technologies Inc. and now belongs to the NVIDIA Corporation. This graphical environment allows for the defining of simulation routines that can use a large set of high fidelity scenarios and a scale display. The rendering engine is based on the Microsoft XNA Framework tool (an API developed by Microsoft for the implementation of PC games and Xbox 360) (XNA Developer Center 2010).

- Interacting with robots using Windows or web-based interfaces. MSRS creates applications that permit the user to monitor and control robots using a web browser. Moreover, the user can send commands to the robot using existing web technologies such as HTML and JavaScript. Furthermore, cameras can be set on robots and controlled from remote locations.

2.4.2.2 Runtime Services

MSRS includes a representational state transfer based on .NET and runtime services that consist of two components:

- Concurrency and coordination (CCR). CCR is a dynamically linked library accessible from any language oriented to .NET Common Language Runtime.

It simplifies the writing of code to handle asynchronous inputs from multiple sensors and outputs to actuators and motors. In the handling of asynchronous operations, it takes advantage of parallel hardware. CCR is appropriate for an application model that separates the different components into segments that can only interact via messages.

- Decentralized software services (DSS). This is placed above CCR and it provides a state-oriented model of services that combines the concept of representation state transfer with a system based on levels and focused on building scalable applications. In DSS, services are considered accessible resources through programming and through the manipulation of the user interface. The DSS application model simplifies the access and responds to the state of a robot using a web browser or an application based on Windows.
- Reuse of modular services using a composite model (Lopez de Toro C and Ribas Xirgo 2008). The use of simple components in the development of high-level functions foresees the reuse of code modules, as well as better reliability and replaceability. Services are considered basic blocks to program applications using MSRS. Services can be used to represent, among other things:

  – Hardware components: sensors, actuators.
  – Software components: user interface, storage.
  – Added components: fusion sensors.

2.4.2.3 Scalable and Extensible Platform

The programming model of MSRS can be applied to a wide variety of robotic platforms, enabling the user to transfer his or her achievements across multiple platforms. The programming interface can be used to develop applications in one or more processor cores. The scalability of MSRS can be seen in that (Microsoft Robotics 2010):

- It can easily expand its functionality. MSRS's functionality can be extended through libraries and additional services. Software and hardware manufacturers can easily make their products compatible with MSRS.
- It supports mixed applications. Remote scenarios allow the connection, from a PC to a robot through Ethernet, Bluetooth, 802.11 (Wi-Fi) or RF (Radio Frequency). Programs can be implemented on PC-based robots that run under Windows operating systems, allowing fully autonomous operations.
- It allows the use of a wide range of programming languages. MSRS robotic applications can be developed using a selection of programming languages, including those used in Microsoft Visual Studio and Microsoft Visual Studio Express (C# and VB.NET) and languages such as Microsoft Iron Python. It is also possible to use other languages that support the architecture based on services using MSRS.

Table 2.3 Comparison of robots implemented in MSRS and Player/Stage

MSRS	Player/Stage
Aldebaran Robotics NAO	–
–	Acroname's Garcia
CoroWare CoroBot	–
–	Botrics's Obot d100
Lego Mindstorms NXT and RCX	–
–	Evolution Robotics ER1 and ERSDK
–	K-Team's Robotics Extension Board (REB) Kameleon 376BC and Khephera
iRobot Create	RWI/iRobot based on RFLEX
KUKA Robotics	–
Parallax Boe-Bot	–
Not present	MobileRobots PSOS/P2OS/AROS
Robosoft's robots	–
–	Nomadics NOMAD200
–	UPenn GRASP's Clodbuster
Segway RMP	Segway's Robotic Mobility Platform (RMP)
RoombaDevTools	iRobot's Roomba
WowWee RoboSapien	–

2.4.3 Differences with Player/Stage

Although Player/Stage is an open platform, with all the advantages that entails customization, MSRS has a number of advantages over it (Lopez de Toro C and Ribas Xirgo 2008):

- The user interface is friendlier thanks to the VPL.
- The use of PhysX in VSE gives greater quality in simulation and a more realistic environment than Gazebo does.
- It has a greater number of modules, which can be added to robots.
- The number of robots implemented in MSRS is greater than that in Player/Stage (Table 2.3).

2.5 Webots

Webots (Michel 1998) is defined as robot simulation software that provides an environment for modeling, programming and simulating mobile robot prototypes. It was developed over a period of 7 years by Cyberbotics Ltd., a spin-off from the Swiss Federal Institute of Technology in Lausanne founded in 1998.

Webots has a complete set of libraries that allows the easy transference of control programs to a large number of real robots in the market (Michel 2004).

Webots can define and modify the whole configuration of a mobile robot, or even of a set of different robots acting in the same environment. Moreover, for each part of the robot it can set properties such as color, texture, weight, layers, and so on. Webots allows the simulation of robots with a wide variety of actuators and sensors.

Webots runs under different operating systems such as Windows, Linux and Mac OSX and is aimed at researchers and teachers interested in mobile robotics. Currently, there are several product licenses:

- PRO: This is focused on RandD, has a custom set of physical objects and monitors their capacity using a fast simulation mode.
- EDU: This is less powerful and cheaper than the previous version is and focuses on the educational field.

2.5.1 General Features

Webots has some relevant features that make it a powerful and easy to use simulation tool:

- It models and simulates every mobile robot, including wheeled, articulated with legs and even flying robots.
- It includes a complete library of sensors and actuators.
- It allows the programming of robots in many languages such as C, C++ and Java, or even from other kinds of software through TCP/IP.
- Controllers can be transferred to real mobile robots, including Aibo, Lego Mindstorms, Khepera, Koala and Hemisson.
- It uses the ODE library to simulate physical behaviors.
- It allows video capture of the simulations in AVI or MPEG format for public or web presentations. It includes several examples of source code drivers and commercial robot models.
- It can simulate multi-agent systems that simplify global and local communication.

2.5.1.1 Robot and Environment Editor

Webots provides a list of sensors that can be installed on the robot and adjusted individually (range, noise, response, quality of vision, etc.). This library of sensors includes:

- Distance sensors (infrared and ultrasound)
- Rangefinders
- Light Sensors
- Touch sensors
- Global position sensors (GPS)

Fig. 2.19 Humanoid robots modeled for Webots (Michel 2004)

- Inclinometers
- Compass
- Cameras (1D, 2D, color, black and white)
- Receivers (Radio, IR)
- Position sensors for servo-motors
- Incremental encoders for wheels

Similarly, Webots also has a library of actuators, including:

- Differential for wheels
- Independent motors for each wheel
- Servo-motors (for the arms, legs, etc.)
- LEDs
- Transmitters (Radio and IR)

Webots also allows the creation of a custom robotic model and has a simple system for the creation of complex simulation environments. It is able to create light flashes, smoke or texture mapping thanks to the use of OpenGL for hardware acceleration. Moreover, Webots can import 3D models from 3D modeling software that follow the VRML97 standard (Vajta and Juhasz 2005). For the simulation of robots with multiple complex joints or camera systems (see Fig. 2.19), this feature makes it a suitable tool for humanoid robots.

2.5.1.2 Realistic Simulation

The simulation system used in Webots is based on virtual time. This makes simulations much faster than tests performed with a real robot (Hayes et al. 2003). Depending on the complexity of the robot's configuration and the benefits of the computer, simulations can be up to 300 times faster than a real robot when using the fast mode. Given the need to conduct a detailed study about the robot's way of working, Webots implements a step-by-step mode.

The robots with joints require accurate physical simulation. Because of this, Webots is based on ODE to perform accurate physical simulations (Mojon 2004). For each robot component, it is possible to specify:

- The mass distribution matrix.
- The coefficients of kinetic and static friction.
- The coefficient of elasticity.

To place more emphasis on the realism of the simulation, each component is linked to the coordinates used for collision detection; devices with servo-motors are controlled by programming the position or velocity. The user can interact with the simulation while it is running using the mouse to change the vision of the environment or to move and rotate objects.

2.5.1.3 Programming Interface

Webots allows programming the robot in C as shown in Fig. 2.20; likewise, it also includes a Java programming interface.

Moreover, every Webots controller can be connected to third-party software such as Matlab, LabView or Lisp through a TCP/IP interface. Webots also adds the possibility of implementing a supervisor program to simulations that require long computational time when the simulation needs the evaluation of many parameters (Cyberbotics 2009).

2.5.1.4 Transference to Real Robots

Webots allows the transference of the control code from real to simulated robots:

- Khepera and Koala: This has a C cross-compiler of Webots controllers and remote controls in every programming language.
- Hemisson: This has a finite state automaton, graphically programmed for the remote control and autonomous modes of execution.
- LEGO Mindstorms: This has a cross-compiler for RCX Java of Webots controllers based on LeJOS. Aibo: It has a C/C++ cross-compiler for Webots controllers based on Open-R SDK.

```
#include "robots.h"
#define BORDER 2000
inline shoot(int dir,int range)
{
    if (range > 200 && range <= 7000)
        Cannon (dir,range);
}
main(){
    int sdir=0;
    int dir=0;
    int range;
    int hadfix=0;
    int cx,cy;
    drive(dir,100);
    while (1){
        int tdir=dir;
        cx = loc_x();
        cy = loc_y();
        if (cx > 10000-BORDER)
            if (cy < 10000-BORDER)
                tdir = 90;
            else
                tdir = 270;
        else if (cy < BORDER)
            if (cy < BORDER)
                tdir = 0;
            else
                tdir = 270;
        else if (cy > 10000-BORDER)
            tdir = 180;
        else if (cy < BORDER)
            tdir = 0;
        if (!speed() || dir != tdir)
            drive(dir=tdir,100);
        if ((range=scan(sdir,10))){
            shoot(sdir,range);
            hadfix=1;
        }
        else if (hadfix){
            sdir += 40;
            hadfix=0;
        }
        else
            dir -=20;
    }
}
```

Fig. 2.20 C code to program a robot (Rognlie 1995)

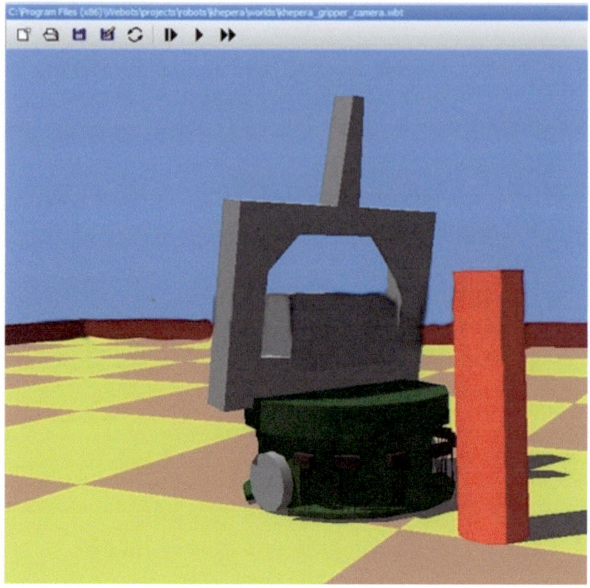

Fig. 2.21 Example of a simulation of a Khepera robot with Webots (Zlajpah 2008)

2.5.2 *Implementation*

2.5.2.1 Khepera

The Khepera mobile robot, produced by the Swiss Federal Institute of Technology in Lausanne, is characterized by its small size and its use in teaching and development processes. Wang et al. (2000) applied Khepera to an environment to create Webots (see Fig. 2.21). The features of Webots that led Wang to use it with the Khepera robot are described below:

- The simulation program can be easily transferred to real robots.
- Its use in previous simulations in Khepera (in the fields of autonomous systems, intelligent robots, evolutionary robotics, machine learning, computer vision and artificial intelligence).
- Both the real and the simulated robots can be programmed in C using the same Khepera API, making the driver source code compatible between the simulator and real robot.
- Webots is also assessed in the use of any other programming language (MATLAB, Lisp, Java) and in the modeling of 1D and 2D cameras and the design of 3D environments using OpenGL.

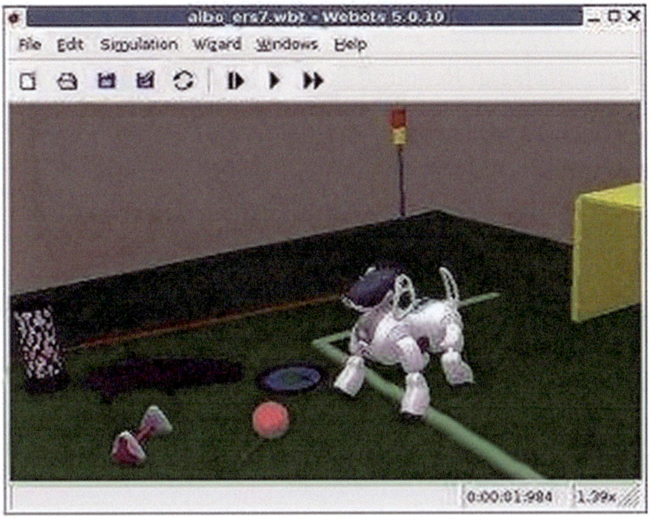

Fig. 2.22 Aibo simulation in Webots

Wang concluded that the possibilities for developing new modules for the Khepera robot and its inclusion in the Webots simulator make it one of the most powerful tools to assist in the RandD of mobile robotics.

2.5.2.2 Aibo

Aibo is a quadruped robot shaped like a dog developed by Sony. Hohl et al. (2006) compared the data obtained when modeling the Aibo robot in Webots with data from the real robot. To do so, they developed a controller in Webots to remotely control the robot. The process for the experiment was as follows:

- A simulation of the Aibo robot in Webots was made by using the official data model, which specified its appearance, kinematic structure, dynamic properties and way of controlling (Fig. 2.22). However, some sensors could not be simulated in Webots because they did not correspond to the nodes described in it.
- After creating a graphical user interface, a robot controller was programmed in Webots. The language used was C to facilitate the work when shifting the source code to Aibo. During implementation, it was observed that the object representing the sensor simulation represented the actual sensor in Aibo. As a result, the call to the read function in the sensor returned the actual value.
- In cross-compiling, using C++ to write programs for OPEN-R and C for the Webots controller facilitated the combination of program files from the Webots controller with the source files that define the specific functions of Aibo.

2.5.2.3 NAO

The NAO (from Aldebaran Robotics) is often present in the RoboCup (RoboCup 2010). Certo (2009) made a model and a simulation of NAO in Webots. To make the simulation, he took into account the rules of the RoboCup, where every game has two types of controls:

- Soccer Player: This is the program that controls the robots.
- Supervisor: This is the program that controls the game.

Webots has the ability to save simulations in .motion files by applying Motion Manager. Thus, the strategies and movements of the robot can be set before the game.

Chapter 3
Service Robotics

Ignacio González Alonso

Abstract In this chapter, a classification of service robotics technology within the digital home is established. This is followed by several examples of the different categories of service robots. The classifications include vacuuming and cleaning, gardening and lawnmowers, personal robotic assistants, telepresence, teleassistance and health, entertainment, home security and privacy and robotic learning categories. Some of these are analyzed and their parts described using SysML formal, open and standard notation. Finally, a brief note about the synergies between professional service robots and home service robots is included.

3.1 Introduction

The integration of computer and human activities has allowed computer science to push the boundaries of technology and today there are more than 6.5 million integrated units in use worldwide in 2007. This figure, as seen in 2008 edition of World Robotics report, and is estimated to rise to 18 million units by 2014 (Gomez 2008). Accordingly, the subset of home service robotics is expected to have 18 million robots by 2011 (IFR/WorldRobotic 2008).

The development of service robots (Schraft and Schmierer 2000) started in the late 1950s and early 1960s with the first industrial robot known as Unimate designed by George Devolop and Joe Engelberger (Mellon Carnegie 2010). Joe also designed Unimation and was the first to market with this machine, earning him the title of "Father of Robotics." By the 1980s, modern industrial arms had already increased their skills and performances through microcontrollers and modern programming languages.

I.G. Alonso (✉)
University of Oviedo, Oviedo, Spain
e-mail: gonzalezaloignacio@uniovi.es

I.G. Alonso et al., *Service Robotics within the Digital Home*, Intelligent Systems, Control and Automation: Science and Engineering 53, DOI 10.1007/978-94-007-1491-5_3,
© Springer Science+Business Media B.V. 2011

These advances were achieved thanks to large investments in automotive companies. Since its beginnings, robotics has been limited to very small and isolated areas apart from the automotive industry, such as the defense or space sectors. In the past decade, because of the economic boom of the 1990s, this has extended into some of the fastest growing fields such as aviation and pharmaceuticals (Barrientos 2002).

With its tremendous growth and widespread use in most prosperous sectors, the production of robots has been optimized, resulting in a significant reduction of costs. It is encouraging that robots have extended to several sectors: construction, agriculture, tourism and ITC among others. A wide range of robots aimed at this particular area is also being developed and marketed at the same time as the industrial ones.

Expectations are high because of a number of factors. These include a robust public acceptance of the first commercial robots, the wide acceptance of IT in general, and familiarity of the population with robots in the workplace, along with a sufficient level of technology at an affordable price. All the evidence implies that robots will become common, and having multiple robots in every home will be as frequent as finding several computers in the same house.

Service robots do not have a precise definition; however, the International Federation of Robotics (IFR International Federation of Robotics 2010) decided to define them as "*robots that work in an autonomous or semiautonomous way to develop useful services, oriented to the well-being of humans and work teams, excluding the repetitive or tedious tasks.*" The International Service Robot Association (Pransky 1996) defines service robots as "*machines that interact and think with the objective of increasing the abilities of the human being and his productivity.*" Both definitions have some intrinsic ambiguity, but are the best found in the current literature. For instance, could an industrial robotic arm be considered a service robot? It is an open question to be answered by applying the definition of a robotic platform rather than from only a technological perspective.

Kawamura et al. (1996) preferred to define service robots as "*sensor-based mechatronic devices that perform a useful service in the activities of humans.*" According to this definition, service robotics stand somewhere between industrial robots and space robots.

The EUROP (European Robotics Technology Platform) (Wendel and Bischoff 2009) distinguishes five areas of application to classify the different types of robots:

- Industrial: Work, partner and logistic robots.
- Professional Service: Work, collaborator, logistic, monitoring, exploration and education robots.
- Domestic Services: Staff, logistic, monitoring and education robots.
- Security: Staff, logistic, monitoring and exploration robots.
- Space: Work, collaborators, logistic and exploration robots.

Currently, robots are being developed for most human environments, and they are becoming generally available because of price reductions. Therefore, the new

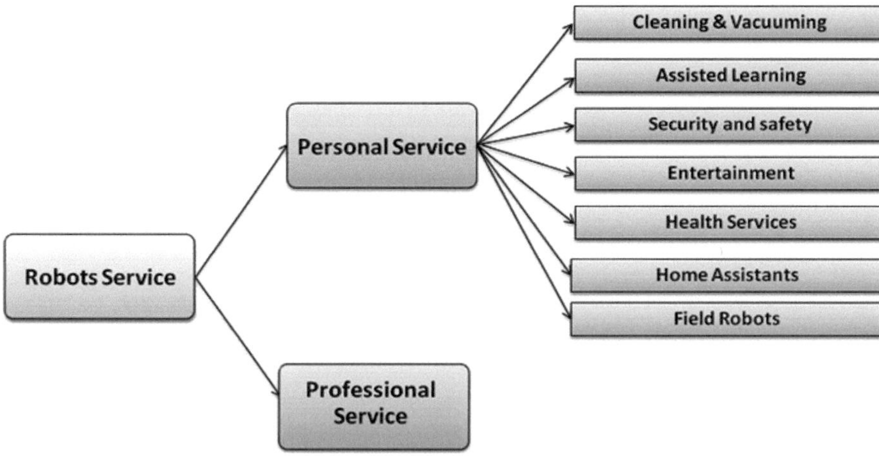

Fig. 3.1 Home service robotics – body of knowledge

category of personal service robotics can be analyzed. The following subsections categorize them and give the most representative examples based on the market, the science and the author's subjective criteria (Fig. 3.1).

3.2 Personal Service

3.2.1 Cleaning and Vacuuming

The main part of the personal robotics market share is held by the vacuuming robots of IRobot. They are the best example of how an application can push the limits of manufacturing and devise a robotics solution to a common problem. The following examples show three different approaches to the same problem.

3.2.1.1 Roomba and Scooba

The Roomba series is the biggest product line for robotic vacuuming within a private home (Jones 2006). Its robots, from series 500 (IRobot 2010), are focused on the house environment and are the best selling service robots with approximately seven million units sold.[1] They have become more and more autonomous over

[1] This figure has not been verified by the manufacturer, but the data were gathered directly from a presentation from the manufacturer.

Fig. 3.2 IRobot Roomba 563

Fig. 3.3 IRobot Scooba

recent years, and with more characteristics. Nowadays, IRobot also has the 600 series for inmotic and industrial environments.

These robots clean up to four rooms in a pseudorandom way, which is an effective compromise between cost and reliance on artificial intelligence. Figures 3.2 and 3.3 show a Roomba 563 Pet series and the Scooba (the floor cleaning version of the Roomba).

ICreate has also been developed for the US market, which is the research platform for IRobot (Fig. 3.4).

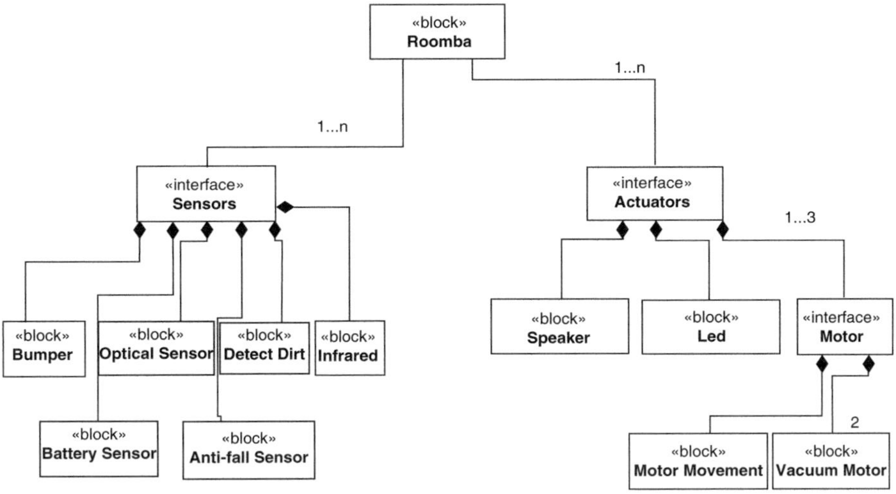

Fig. 3.4 IRobot Roomba SysML block diagram

Notes About Development with Roomba Robots for the Digital Home

I Communication protocol
Given the advantages offered by UPnP (Santana 2005), this was the protocol chosen for service robots such as Roomba. UPnP works with an architecture that provides point-to-point connectivity to give users the possibility of automatically obtaining dynamic IP addresses.

II Microsoft Robotics Developer Studio and Player/Stage in the IRobot Roomba case
The processes performed by robots require constant interaction between software and hardware elements, so it is often necessary to combine the knowledge of both to carry out the development. This can be achieved by using simulation techniques; Microsoft Robotics Developer Studio and Player/Stage are both open platforms (one free but not open source in the concept definition) that can help with the task. They provide Roomba robots with two platforms for control and simulation.

3.2.1.2 Mint Automatic Floor Cleaner

As stated in the above subsection, the navigation algorithm for the Roomba series is pseudorandom. To avoid unclean spots and to reduce operation time, new products have been arriving in the market with improved navigation algorithms and systems. Mint (Mint Evolution 2010) is one of the two that will be described in this section.

Mint's navigation algorithm comes from NorthStar navigation (North Star Evolution Robotics 2010), which makes this robot a more efficient navigation platform than the IRobot alternatives. However, it also mops so it could fit better the expectations of the final user if he or she has a wooden floor (Fig. 3.5).

Fig. 3.5 Mint

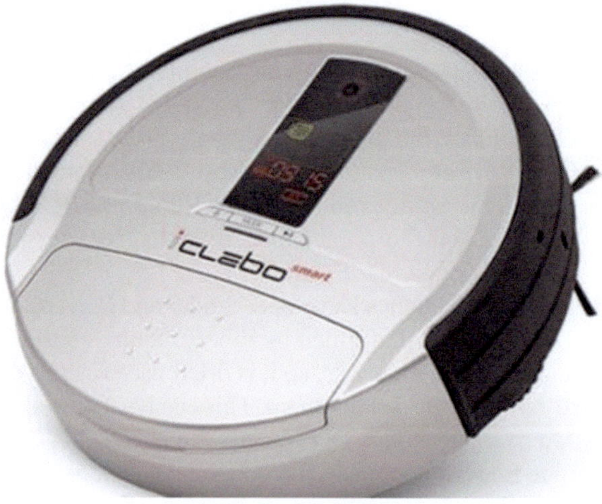

Fig. 3.6 IClebo smart

3.2.1.3 IClebo Cleaning Solutions

Another option in the competitive market of vacuuming solutions is from Yujin Robotics (Yujin Robot 2010). With one low cost and a different intelligent solution, the IClebo platforms are viable alternatives for vacuum cleaning.

The IClebo Home is not that different from the IRobot technologies, but the IClebo Smart robot is really a more powerful alternative to the IRobot and Mint combined. The disadvantages of these products are their size and weight but depending on the use, it would be a good idea to have a big dirt deposit chamber (Fig. 3.6).

Fig. 3.7 Aquabot Pool Rover

3.2.1.4 Pool Cleaners

Another interesting application for automatic cleaning is the swimming pool cleaning. Different manufacturers compete in this segment using contrasting approaches. The Aquabot (Aquabot 2010) manufacturer is another alternative with its Aquabot Pool Rover (Fig. 3.7).

3.3 Green, Agricultural and Lawnmowing

The second most popular application for home robotics is automatic lawn mowing. Like vacuum robots, lawnmower robots focus their services on a specific application. The robotic platform infrastructure behind it is similar to that of vacuuming systems, but it is designed for outdoor work, as can be seen by comparing the Roomba SysML block diagram and the LawnBot SysML diagram. The presence of markers to help the robot fix its working area was initially its main limitation. Modern robots are now more context-aware, and their sensors and localization systems allow them to autonomously navigate inside their working areas. Common to all solutions in this subset is the presence of safety systems to avoid people or animals (Figs. 3.8 and 3.9).

Despite the large and growing markets for service robotics in homes and smart cities, enormous research continues on autonomous machinery for applications to agro-farming tasks.

3.3.1 Green Botics

IRobot and other manufacturers are starting to consider efficient energy management as an important feature in their products. The DH Compliant standard set by

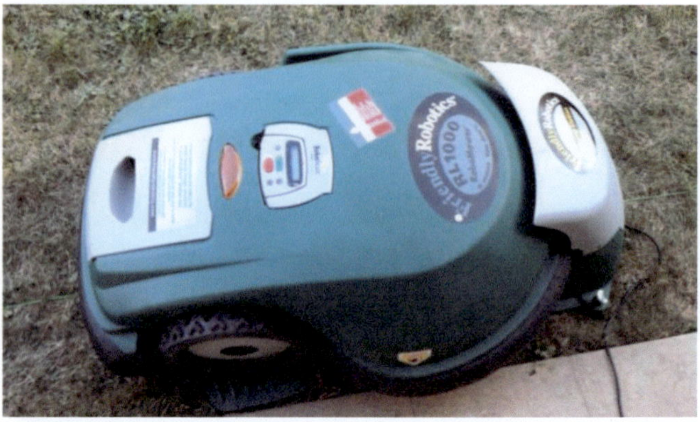

Fig. 3.8 Friendly Robotics LawnBot (Robomow 2010)

Fig. 3.9 KA LawnBot (LawnBott 2010)

its robot manufacturer's consortium, similar to efforts seen in Europe, is also being applied to the same problem and finding the same solutions. The challenge is huge, but it is a must for any manufacturer seeking an opportunity in the modern robotic home environment.

3.4 Home Personal Robotic Assistants

Although they are not yet in the mainstream, different applications for home personal robots are resulting from the research efforts of a number of private and public initiatives. As with any type of social skills (Breazeal 2004), robotic platforms must try to interact with humans in unstructured environments, and they have done so with some success. These kinds of platforms are Human-shaped and are being developed to ultimately replace the human majordomo or maid. They have also developed some interoperability software services to equip any home robotic platform with some intelligence, energy management and localization systems according to the DH Compliant protocol. The following robotic platforms are serious attempts to achieve that final goal.

3.4.1 Examples

Personal Robot from Willow Garage, the PR2 (Cousins 2010; Willow Garage 2010) is a robotic platform for research and education. Its main characteristic is the use of ROS (the open source operative system, which is a fork from a Linux distribution with real time and other robotic framework utilities). It has been used for automatic plugs into electric walls as well as for experimenting in different unstructured scenarios present in a house (Fig. 3.10).

In the near future, linked to a price reduction, it may be possible to find another interesting alternatives for home personal robotics. The FutureBot (Futurerobot 2010), a restaurant and museum assistants from Futurerobot (Fig. 3.11).

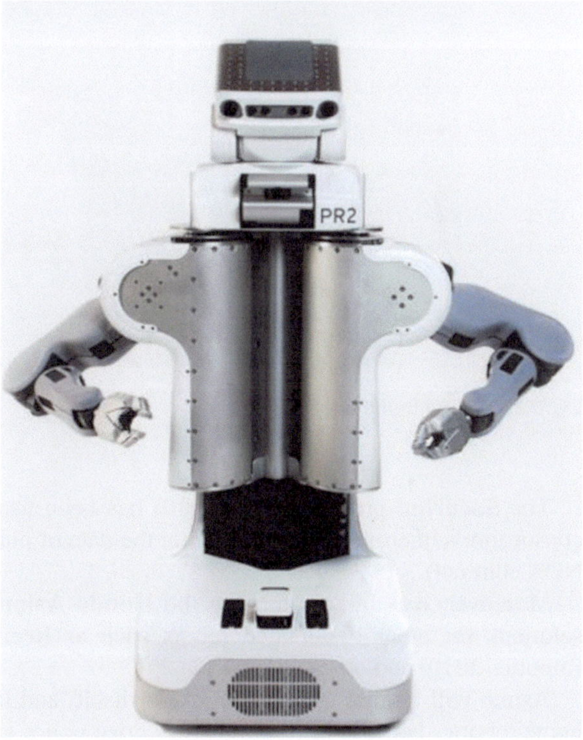

Fig. 3.10 PR2 from Willow Garage

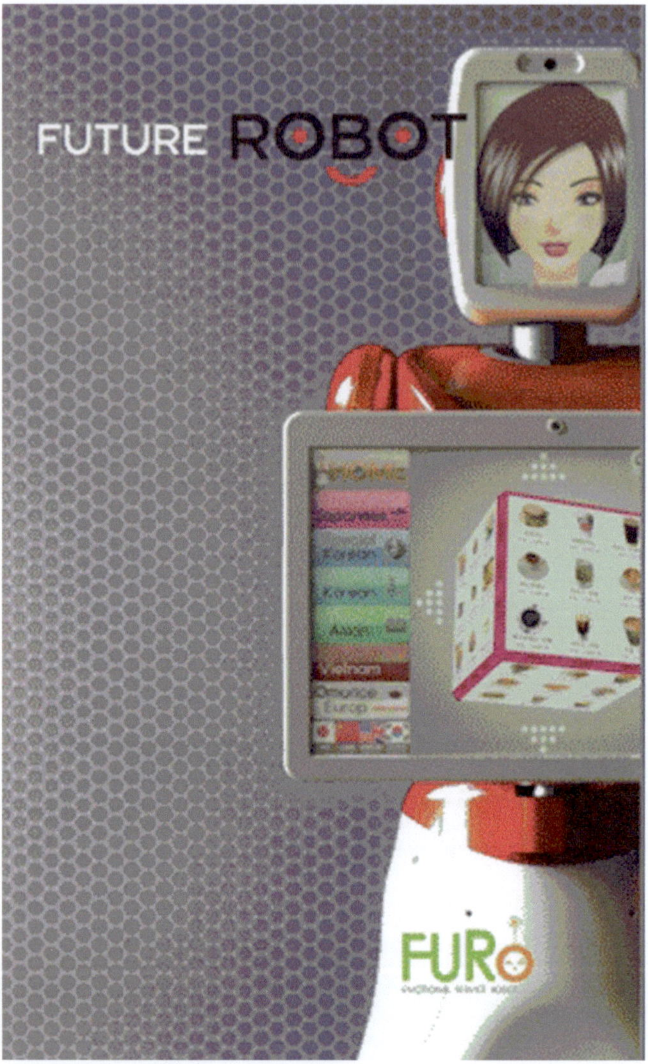

Fig. 3.11 FURO from Futurerobot

The **Sacarino** project (Cartif 2010) has been used inside the DH Compliant consortium's interoperability efforts (at the date of publication still protected by an NDA contract).

Moreover, it is fair to mention the **Honda Asimo** (Honda 2010) and similar solutions for other two-legged robots such as Reem-B from Pal-Robotics (Pal Robotics 2010) and others (Fig. 3.12).

Asimo will require an entire book for itself, and it does not fit the category of home robotics because of its company's price policy (Fig. 3.13).

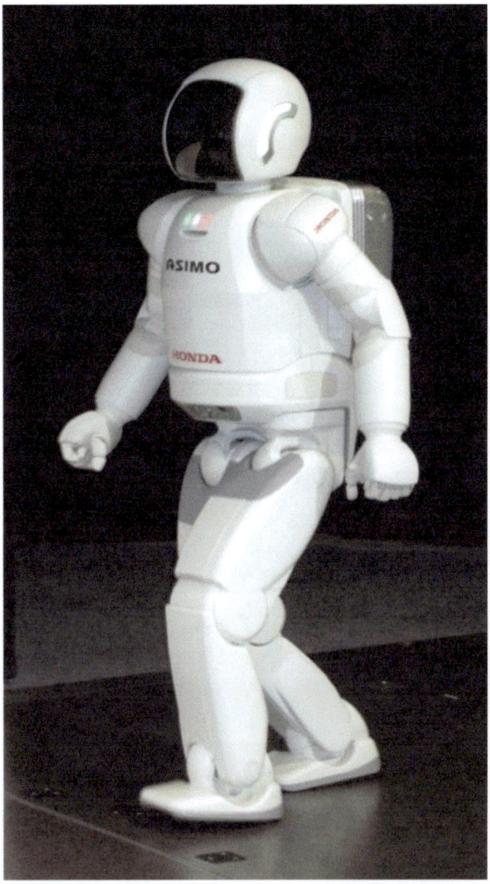

Fig. 3.12 Honda Asimo (Sakagami et al. 2002)

3.4.2 Home Robotics Interoperability: DH Compliant Services

Software services in a network environment have their own role in house robotic assistants. Digital Home Compliant is an initiative led by Ingenium, University of Oviedo, Domótica Davinci, University of Sevilla, Cartif and Movirobotics that aims to achieve an interoperability virtual device based on DHC-Protocol and UPnP. It has several services to help a robot interoperate with other robots and with home and building automation services. For instance, DHC-groups (for house services cooperation), DHC-Energy (for green energy management), DHC-Intelligence (for business rules developments and machine learning), DHC-Localization (to get the position of a device) and DHC-Security&Privacy (to help the user in managing its privacy and protecting its home).

The architecture of DHC is depicted in Fig. 3.14.

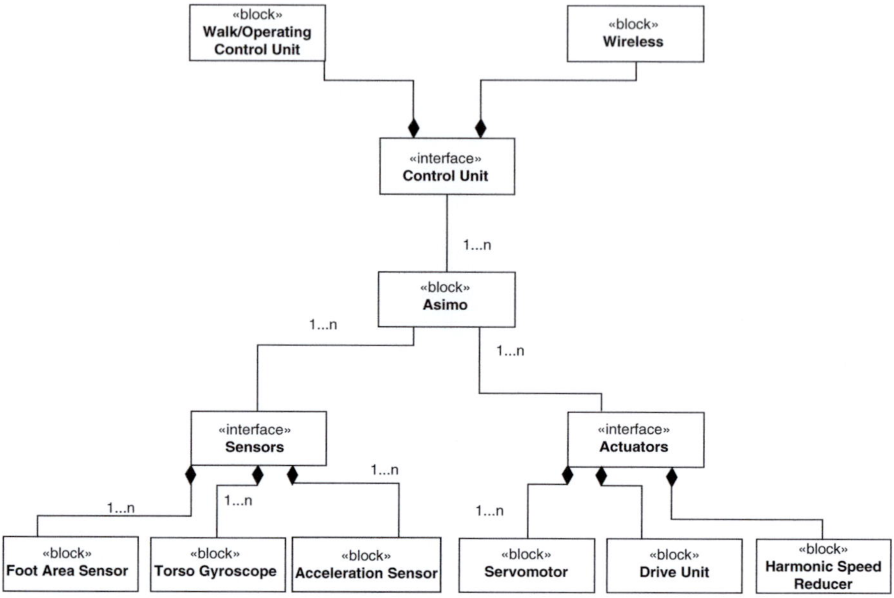

Fig. 3.13 Asimo SysML block diagram

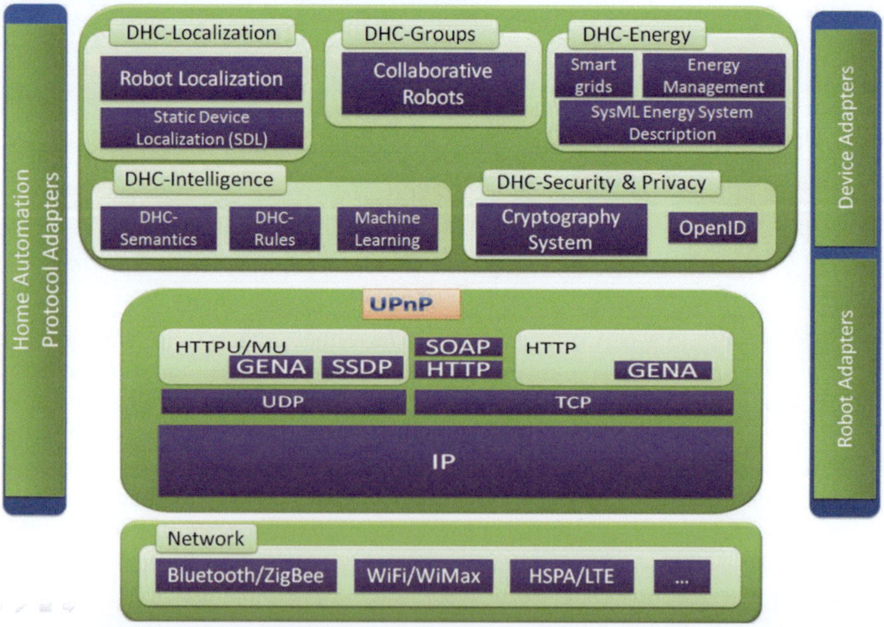

Fig. 3.14 DH Compliant architecture

3.5 Telepresence, Teleassistance and Robotic Health Services

The concept of telepresence (Graf et al. 2004) has been widely developed by modern research and market applications. It reflects the idea of being in two places at the same time, and mixes the power of ITC with robotic mobility, helping the owner sense and act remotely through the teleoperation of these platforms through a network (nowadays that means the Internet). Three examples of these platforms follow.

The most cost effective option for telepresence is the Rovio Wow-Wee (Begum et al. 2010; WowWee 2010). This is a cheap alternative, but incorporates an indoor navigation system based on an infrared vision recognition system. Its robot wheels are multidirectional and it can be compared with a 4 × 4 in the automotive sector. Its main weak point is that it does not have a camera at the height of a human face. However, its moving camera has tried to satisfy that need (Fig. 3.15).

Despite the price, it is of interest to analyze the Anybots QB telepresence solution (Fig. 3.16).

The final alternative shown is the Rovio competitor, the Spykee (Spykee World 2010) (Figs. 3.17 and 3.18).

3.6 Entertainment

Automatisms were already available when robots started to be used as entertainment tools (Karakuri 2010) or to perform magic tricks. Modern entertainment robotics applications for the digital home are linked with robotic toys. Moreover,

Fig. 3.15 Rovio Wow-Wee

Fig. 3.16 Anybots QB

Fig. 3.17 Spykee

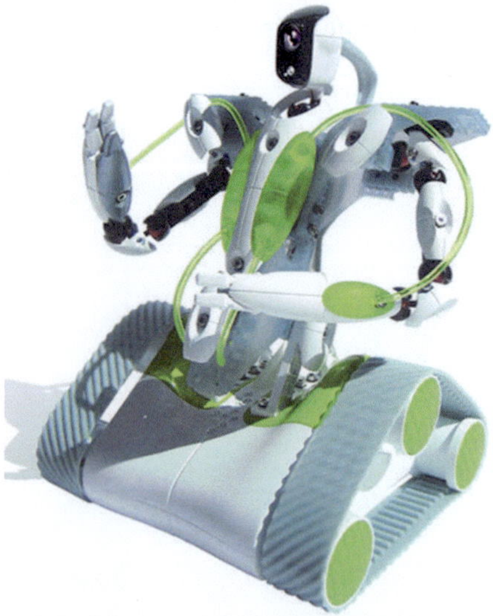

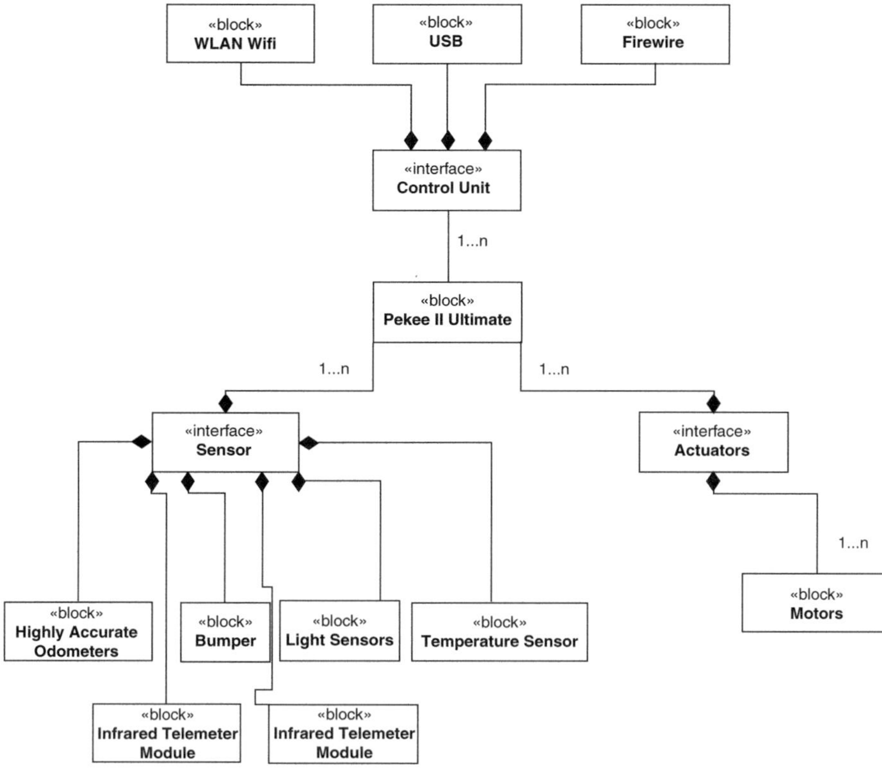

Fig. 3.18 Spykee SysML Block diagram

they have developed platforms intelligent enough to interact with their owners (Kerstin et al. 2003). They have also developed kits for adults who discovered that they enjoyed learning how to use them and how to build variations of their own. Of course, there are other entertainment robots, such as those for shopping malls, theaters and so on. But they are outside the scope of this book.

3.6.1 Playing

There is an enormous variety of robotic toys presenting many benefits and innovative approaches in amusement and assistive technology for children with autism. Failures in this area will help other robotic toy developers understand the fact that this market has not only typical high-technological device constraints, but also limited budget constraints for customers.

Fig. 3.19 Wow-Wee Tribot (Marco et al. 2010)

3.6.1.1 Tribot

From the WowWee (WowWee 2010) company, this robot explores several social interactions with children and is a good platform to introduce elementary programming concepts to youngsters. Its remote control is the perfect excuse to show how to program the device with a highly intuitive interface (Fig. 3.19).

3.6.1.2 Robotic Teddy Bear

From the MIT Media Lab, the Teddy Bear (Matsumaru 2009) is a platform for developing health solutions (MIT media lab 2010). Examples of those health applications of that robot are autism treatment or old people care. Of course, the research in personal health appliance should have also synergies with those robots.

3.6.1.3 Pleo

> "In all science, error precedes the truth, and it is better it should go first than last" Hugh Walpole

Technicians who develop robots sometimes make mistakes, just like any other human. One of the most frequent mistakes is losing focus on solving market needs, confusing the fulfillment of a customer's expressed desires with the beauty of a particular technique. While understanding the ease of humanizing a

Fig. 3.20 Pleo robot toy

Fig. 3.21 Boe Bot

device, one can lose sight of the fact that developing a product without clearly understanding the needs of the customers could lead to failure. The same principle applies to not understanding that a $600 toy is not a toy; no matter what the Pleo functions were. The Pleo toy had those problems; and consequently, it failed (Fig. 3.20).

3.6.2 *Robotic Kits*

This section explores some robotics kits that are having enormous success in South Korea in terms of developing new applications for sumo or martial arts robot games (Robobuilder 2010). Examples of these kits include Robotis and RoboBuilder (Figs. 3.21 and 3.22).

Fig. 3.22 RoboBuilder

3.7 Security and Safety Robotic Services

José Luis Rubio Pérez
Movirobotics Albacete Spain

3.7.1 Home Security

Home security is a problem with no widely accepted solution. Therefore, there are many opportunities to achieve a marketable solution for mainstream sales. It is remarkable that some attempts today in this category can be used inside buildings and homes to provide better protection and surveillance. It is impossible to describe all the security platforms that have been developed for this purpose, so a few examples will be considered as potential future home security devices. The following five potential robotic platforms could play a role in home security (Figs. 3.23–3.28).

It is not only in homes that security robots play a role, but also the defense sector has been a ready market for many solutions. If these become affordable (and they will), they will have an even more expanded role in our smart cities and societies of the future. Examples include Predator (General Atomics 2010), Big Dog (Wooden et al. 2010; Boston Dynamics 2010) and the Samsung surveillance unmanned vehicle (Samsung 2010).

Fig. 3.23 IRobot PackBot

Fig. 3.24 Pointman robot

3.7.2 Privacy Considerations

All technologies have some risks, and service robotics is no different. Therefore, the importance of understanding and managing these questions is crucial. Inside the digital home, a key question is that of privacy and the openness of traditionally protected space in homes.

Fig. 3.25 mSecurit

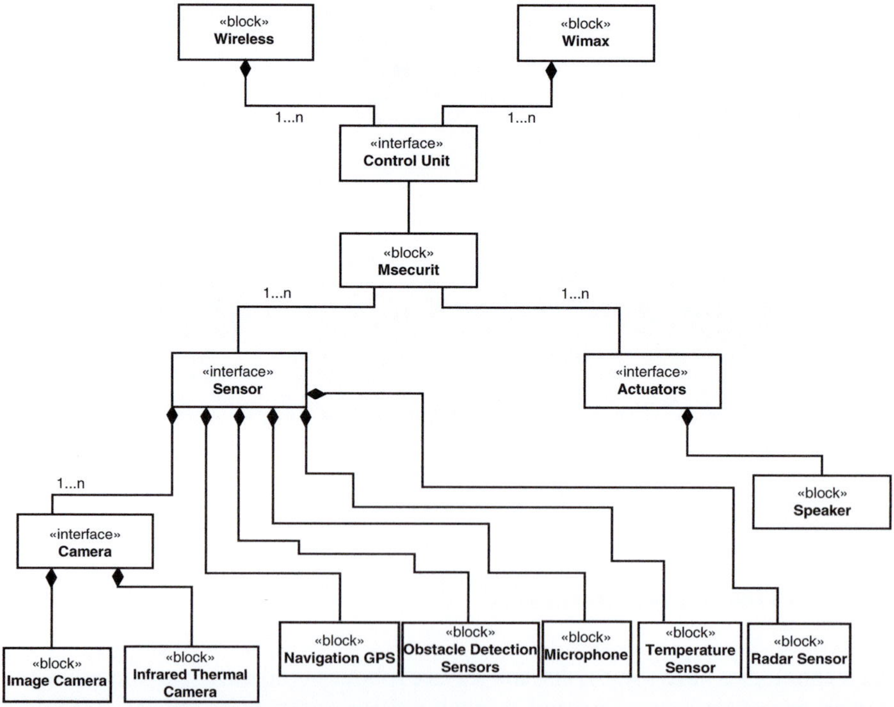

Fig. 3.26 mSecurit SysML block diagram

Fig. 3.27 Canadarm2 (Mamen 2003)

Fig. 3.28 Ultra light unmanned aerial vehicle (UAV)

An interesting approach to this problem can be obtained from a comparison with privacy management inside social networks or O.S. security policies. This might show how the intentions of users have to be prioritized over any other consideration within their own home. The right to have control of your home is not only a primary feature of privacy acts of different countries, but exists in the upper range of laws and norms of any democratic legal system, such as constitutional or root law books.

Therefore, it is of high importance to manage this feature carefully if digital home robots are to be effective.

3.8 Home Robotic Assisted Learning

Learning is a lifetime task, from childhood to adulthood. Science and technology benefit from the visualization and motivation a robot gives students in the areas of physics, mechanics, electronics, computer science, and other similar disciplines. Furthermore, mobile phones connected to our houses and robots allow us to satisfy both our random curiosity and our ongoing learning needs. E-learning is outside the scope of this book, but there exist different robotic software and hardware that represent good examples of robotic learning solutions.

For example, the Lego NXT Mindstorms is the best attempt at a cost effective and powerful platform for robotic facilities. In addition, it has an enormous deployed base so it can be seen as the *de facto* standard in the learning robotics area (Fig. 3.29).

Of course, there are hundreds of problems that admit of robotic solutions, but we would like to mention one more that is only a software platform. This is the suite for robotic programming and simulation from National Instruments Lab View (NI LabVIEW 2010) (Fig. 3.30).

Fig. 3.29 Lego NXT Mindstorms

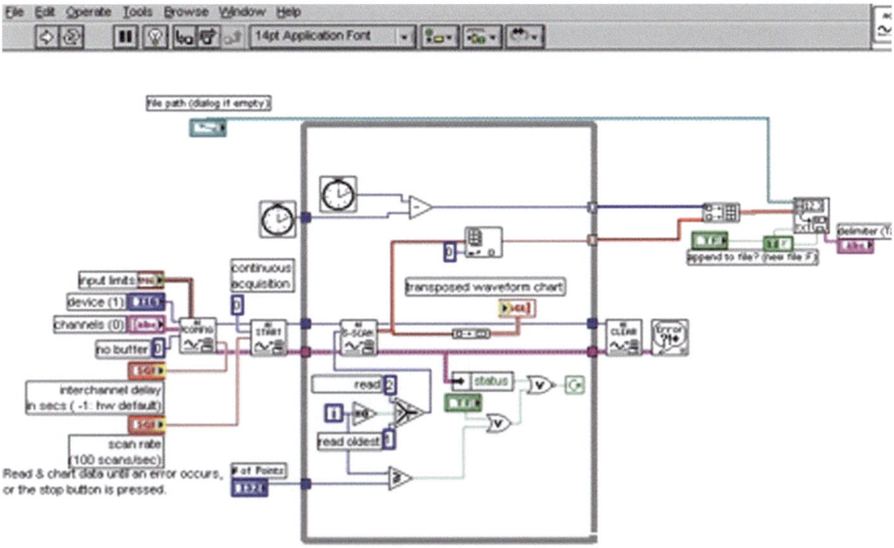

Fig. 3.30 National instruments lab view

3.9 Other Service Robotics in the Professional and Home Environment

The fields of professional service robotics, automated logistics and industrial robots will create enormous synergies with the digital home. Here it will be the same situation as in security robotic solutions. They will be the leaders and the pushers for new technologies that will came to the mainstream when the expected price reduction occurs.

3.9.1 Professional Service Robots

Professional service robotics is the branch of service robotics with the best growth forecast, since the industry is already using solutions similar to home robotics but in a different context. The aim of professional service robotics is to enhance the ability of people to perform tasks required by their jobs. The current professional robotics strategy is to replace or cooperate with the worker on tasks involving high risk or tedious tasks using intelligent robotic co-workers or teleoperated robots under constant human supervision. However, other solutions are being developed to give a greater capacity to activities that demand important physical skills or if operators are disabled. There is no doubt that, in the unstoppable search for the lowest costs and maximum benefit, a more profitable way to use professional robots instead of cheap workers from underdeveloped countries will be found (Fig. 3.31).

Fig. 3.31 Corobot

The main application fields of professional service robotics are the following. All have enormous potential within the digital home robotics context:

- Field and outdoors (agriculture, forestry, mining, etc.);
- Autonomous transport (fleets of vehicles, driver assistance, etc.);
- Professional cleaning and the inspection of different kinds of infrastructures (buildings, ships, pipelines, bridges, etc.);
- Construction and demolition;
- Logistics (logistical tasks in hospitals and offices and the delivery of mail/food/medicine, logistics in workshops, etc.);
- Underwater applications (exploration of the marine fund, inspection and repairing of pipes, etc.);
- Medical robots and rehabilitation; and
- Emergency situations (natural disasters such as fires, earthquakes, floods and human disasters such as bombings, explosions in chemical plants, energy, etc.) (Figs. 3.32 and 3.33).

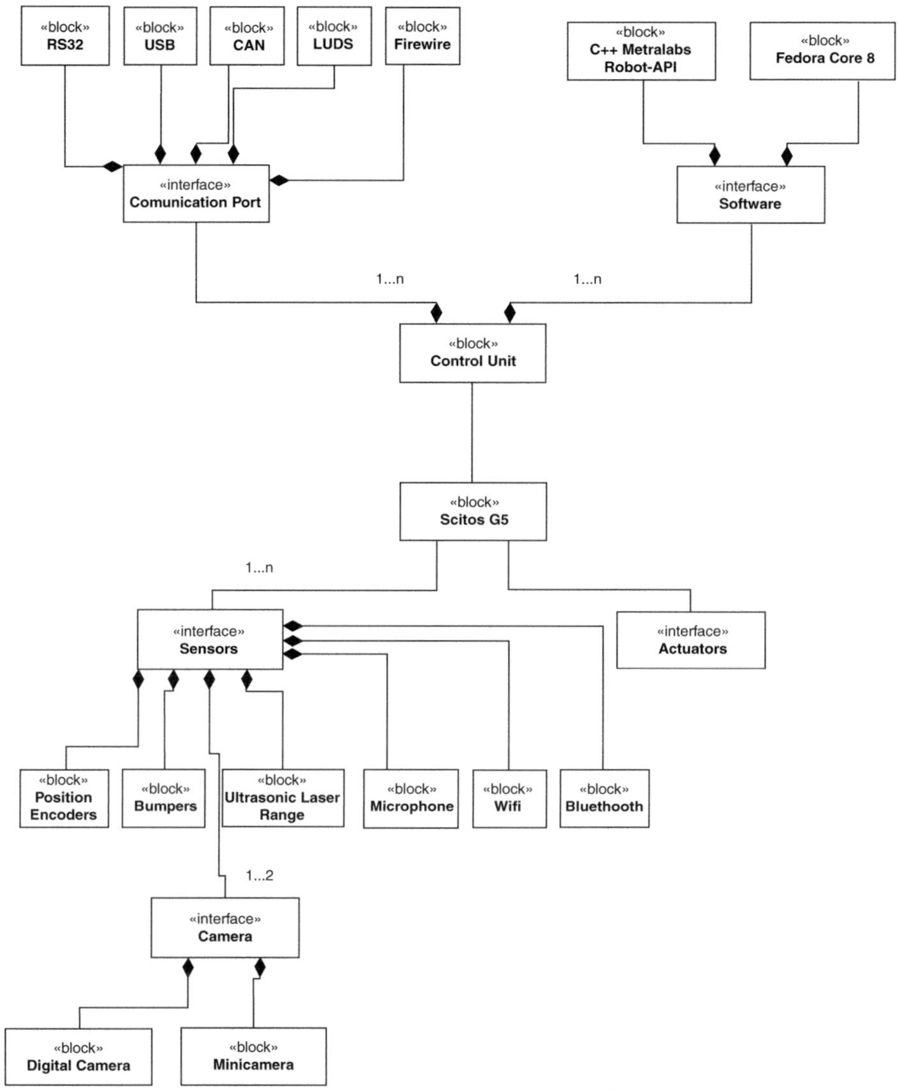

Fig. 3.32 Scitos GT SysML block diagram

Fig. 3.33 RobuCab from Robosoft

Chapter 4
Integration of Service Robots in the Smart Home

Mercedes R. Fernández Alcalá, José M. Maestre, and Javier Ramírez de la Pinta

Abstract In recent decades, the number of robotic standards has increased, and this progress has encouraged the integration of service robots and growth in the number of robotic devices with various communication protocols used in the smart home. In this chapter, we study different standards that could be used for the integration of mobile robots and unmanned vehicles in the digital home. As will be seen, the origins of these standards are twofold. On the one hand, standards have been developed in a military context such as JAUS or 4D/RCS, which is logical given that the control and coordination of autonomous vehicles has many potential applications in this field. On the other hand, standards have been developed in a computer science context, where interoperability between the different agents that may interact in a networked environment is a major problem.

4.1 Introduction

During the early years of computing science, only large organizations such as NASA or the US government could afford to have computers. At that time, no one would have ever imagined the astonishing evolution of computers together with the continuous price drop in consumer electronics. Nowadays, there is at least one computer in almost every home in developed countries, with enough computational power to ridicule the first computer systems. Advances due to the evolution of computers are the first breakthrough in the field of the digital home.

M.R.F. Alcalá
Infobótica Research Group, University of Oviedo, Oviedo, Spain
e-mail: fernandezmercedes@uniovi.es

J.M. Maestre (✉) • J.R. de la Pinta
Department of Systems and Automation Engineering, University of Seville, Seville, Spain
e-mail: pepemaestre@cartuja.us.es; jrdelapinta@cartuja.us.es

I.G. Alonso et al., *Service Robotics within the Digital Home*, Intelligent Systems, Control and Automation: Science and Engineering 53, DOI 10.1007/978-94-007-1491-5_4,

© Springer Science+Business Media B.V. 2011

The next revolution in the smart home is expected to come from the world of robotics. At present, the use of robotics is limited to industrial areas, although service robots that assist us in routine tasks such as cleaning the house, mowing the lawn or even preparing meals are becoming common. Nevertheless, different problems have to be solved before service robots become as popular as computers. In particular, interoperability between the different systems that may exist in future homes is an ongoing issue.

The idea that the reader must have in mind during this chapter is interoperability. Interoperability is the key component to solving the smart home jigsaw puzzle. Thus, in this chapter, we will place special emphasis on the interoperability aspects of the different standards. In addition, different research projects on the interoperability and control of robotic systems and unmanned vehicles will be surveyed. Behind all these standards and projects, there are stories of success and failure, and many valuable lessons about the complex world of interoperability. At this point, it is difficult to know if any of these alternatives will prevail and become a consolidated standard for the integration of robots in the digital home. However, what we know for sure is that any succeeding standard will have learnt from all that will be presented here.

4.2 Military Standards

4.2.1 Joint Architecture for Unmanned Systems (JAUS)

The JAUS standard was developed for the US Defense Department (English 2007) by the JAUS Work Group, which is composed of research groups from the government (US Army ARMDEC), industry (SSC San Diego, WINTEC Inc., iRobot) and academia (Virginia Tech, University of Florida). JAUS was defined as an open and scalable standard that would meet the needs related to the communication of unmanned systems regardless of the platform used. The development of JAUS has tried to meet the following six goals (Wade 2006):

1. Independence of the vehicle's platform;
2. Isolation of the mission;
3. Hardware independence;
4. Independence from the technology;
5. Independence from the operation; and
6. Independence from the connection used.

The JAUS architecture is composed of three levels:

- Level 1 – Inter subsystem: The purpose of this level is to support interoperability between subsystems. It is responsible for specifying requirements between the subsystems (Robot to Robot, Robot to Controller, Controller to Controller).

- Level 2 – Inter nodal: The purpose of this level is to support the interoperability between nodes. To this end, it specifies requirements between the subsystems (interoperability between data loads or between the on-board control and data loads).
- Level 3 – Inter components: The purpose of this level is to provide a reusable software source. It specifies requirements for each component (component by component, such as sensors and motors).

In 2004, a process of transition from the JAUS Work Group to the Society of Automotive Engineers (SAE 2010) started. This developed the standard through the AS-4 (Technical Committee on Unmanned Systems) (SAE 2006). The following norms migrated from JAUS to a framework based on the following services:

- JAUS Transport Standard, AS5669 (SAE-TS 2009). This is in charge of defining the creation of packets with the destination and source addresses and TCP and IP headers and links.
- JAUS Core Service Set, AS5710 (SAE-CSS 2010). This is responsible for providing the means for the software entities in an unmanned system to communicate and coordinate among their activities.
- JAUS Mobility Service Set, AS6009 (SAE-MSS 2009). This is in charge of making the migration from the first drivers to the new development platform of the AS-4.

Today, the main application of JAUS is focused on the use of unmanned civilian and military vehicles.

4.2.1.1 Application of Military Unmanned Vehicles

A major center for development of military unmanned vehicles exists at the SPAWAR Systems Center (SSC) in San Diego (California). There, a JAUS work team focuses on the development of surveillance systems, such as MDARS (Mobile Detection Assessment Response System), which are used in autonomous vehicles for military bases with restricted access.

The US Defense Department uses MDARS to meet security and surveillance needs in hostile environments for humans. In this way, it provides an integrated solution, where unit patrol vehicles are controlled just by a single control operator. Moreover, SSC has developed a distributed processing system called Multiple Resource Host Architecture (Everett et al. 2000) which, along with MDARS, was tested by the JAUS work team in December 2003 to demonstrate the level of interoperability between control operator units (COUs) and unmanned systems (Nguyen 2005). In this experiment, COUs were equipped with a screen capable of displaying the statuses of each patrol vehicle, and thereby they controlled each one of the unmanned systems (Carroll et al. 2004).

These experiments show how the JAUS architecture provides interoperability for the remote control of unmanned systems while fulfilling the objectives mentioned in the general characteristics section.

4.2.1.2 Application of Civil Unmanned Vehicles

In 2004, Virginia Tech launched a project to implement simultaneously the JAUS standard in the following seven unmanned vehicles:

1. MATILDA
   This was the first interoperable vehicle designed by Virginia Tech in 2002. It was designed as an evaluation, development and demonstration platform of the JAUS standard. It had to fulfill some functional requirements:

   - It had to be teleoperable through a COU;
   - It had to be capable of driving autonomously via GPS commanded by a COU;
   - It had to interact with other subsystems of JAUS (either vehicle or COU);
   - It had to accept JAUS workloads from other devices;
   - It had to allow an easy modification and/or addition of intelligent software; and
   - It had to ease the demonstration, evaluation and testing of the JAUS standard.

2. JOHNNY-5
   This was developed in 2004 to participate in the AUVSI Intelligent Ground Vehicle Competition in 2005. Owing to its robustness and its capability to navigate via GPS, it quickly replaced MATILDA. The main problems of this model were the failures in the camera interface and the starting force on the wheels.

3. CADILLAC SRX
   Grant Gothing and Jesse Hurdus, researchers from Virginia Tech, managed to implement the JAUS standard on the Cadillac SRX, creating the first luxury unmanned vehicle in the world (Gothing and Hurdus 2006). The challenge of this model depended on development of a JAUS-based vehicle able to use potential field methods (Koren and Borenstein 1991) for navigation. The result was the creation of a software topology, based on operational subsystems, nodes and components (see Fig. 4.1).

   However, when they launched this vehicle in the Blind Driver competition (Blind Driver Challenge 2010) they detected some issues that could be improved

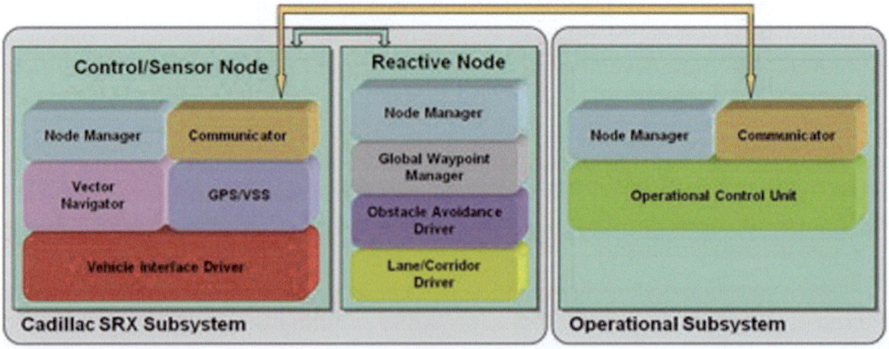

Fig. 4.1 JAUS topology (University of Seville 2010)

(Faruque 2006). For example, every driver had to know the turn angle of the vehicle and, according to the control messages of the JAUS specification, only one controller per component was allowed.

4. GEMINI
   Gemini was developed as an extension of Johnny-5. The idea was to create an articulated robot with four wheels. It won the JAUS Award at the AUVSI Intelligent Ground Vehicle Competition in 2006 because of its refined design, the long life of its batteries (5 h), its innovative mobility and the ability to deal with bigger workloads under the JAUS architecture.
5. HELIUM RED (Unmanned Ground Vehicle; UGV) and THE RMAX (UAV)
   HeLiUm RED (HElicopter LIfted UnManned Reconnaissance and Exploration Drone) redefines the traditional notion of collaboration between UAVs and UGVs (RMAX-HELIUM THE RED). This small unmanned vehicle is light enough to be carried by the VT Yamaha RMAX UAV. Initially, the JAUS standard was implemented to simplify communication with the vehicle; however, vehicles are usually treated as subsystems of the JAUS architecture, but in the project HELIUM RED, the UGV operates as a single node.
6. ROCKY
   This is another example of the vehicles used by Virginia Tech in the DARPA Grand Challenge. The JAUS implementation in Rocky has taken place in two stages:
   - Teleoperability: Through the primitive driver, they could make sure that the vehicle was teleoperated making use of the COU, but nowadays with the use of Global Position/Speed Sensors, the COU, speed and position can be kept on track and transmitted through a connection service.
   - Portability of the basic code from Cadillac SRX directly to Rocky. This feature can be seen as a demonstration of the reusability existing when developing autonomous vehicles under the JAUS architecture.

Owing to these achievements, Virginia Tech established, as functional requirements, that their prototypes had to be interoperable with other JAUS subsystems (applied to both COUs and vehicles). Throughout this research, they realized the need to integrate some specifications in the JAUS Service Specification standard that would make use of messages in charge of waiting for a response that will allow the COU and the vehicles to make behavior decisions for a better interaction between them.

With respect to the development of unmanned vehicles, the company TORC started the ByWire XGV Project (TORC 2010). This project is being developed over a Ford Escape Hybrid using the JAUS standard as a platform to interact with the different parts of the car (steering, throttle, brakes and gear system). The vehicle has an Ethernet interface installed in a central console that allows for remote control of the vehicle by a COU, making use of the SAE AS-4 (JAUS) architecture. The use of the JAUS standard makes sure that ByWire XGV is compatible with any other platform developed on JAUS. It is important to note that the ByWire XGV has maintained speeds of 160 km/h. The DARPA Urban Challenge (DARPA 2007) checks the utility of unmanned vehicles in traffic environments and assesses how they stick to conventional rules of the road. This is a challenge for participants to ensure that unmanned vehicles can perform complex movements such as parking or taking navigational decisions at intersections. In 2005, the DARPA Grand Challenge competition, the University of Florida and Virginia Tech competed with their unmanned vehicle projects based on JAUS.

Applied Research Inc., Virginia Tech, University of Florida, iRobot and the US Air Force Research Lab showed the importance of interoperability in robotics in an experiment (Clark 2005a, b). To this end, each consortium member made their COU able to interact with all robots and control all loads. The benefits of the JAUS standard were successfully proven after showing the independence of the technology used in unmanned vehicles and robots.

Baity (2005), talking about the future of JAUS, mentions the need to focus on development of software. This author says that it is a primary point to take into account to minimize problems in the progress of UGVs.

4.2.2 Other Military Standards

4.2.2.1 4D/RCS (Real-Time Control Systems)

The 4D/RCS architecture provides a reference model for military unmanned vehicles. 4D/RCS is a method of designing, integrating and testing intelligent systems software for vehicles that have a certain degree of autonomy (Albus et al. 2002a). It is an autonomous intelligent control system architecture for vehicles that can be either teleoperated or fully autonomous.

4D/RCS (Kim et al. 2002) specifies the way in which software components are distributed and interconnected, and that is the reason why it became a model for military unmanned vehicles. The importance of this standard lies in the way in which unmanned vehicles must manage situations in hostile environments to complete their missions. As a result of the above features, the 4D/RCS fulfills perfectly the specific needs of the Department of Defense and US Army standards (Albus et al. 2002b).

4D/RCS architecture was based on the assumption that different knowledge representation techniques may offer greater advantages. The aim was to cover all of them to create a real-time control system for objects that move in the real world (Schlenoff et al. 2006).

The Demo III UGB Program (Shoemaker and Bornstein 1998) developed and demonstrated advances in control of unmanned systems, especially small UGVs under supervised control. That is where the 4D/RCS architecture and its characteristics arose. This protocol allows intelligent vehicles to adapt to a changing world, to extract deeper information from a dynamic world and to merge such information with previously available information to improve a vehicle's performance.

The intelligent control of a 4D/RCS system is based on three layers of abstraction:

- A conceptual framework. This is the highest layer of abstraction and covers the full range of operations that involve intelligent vehicles, from a simple actuator for some milliseconds to lots of vehicles during long periods of time.
- A reference model architecture. This defines a hierarchical control structure and at each level functional processes are included.
- Engineering guidelines. These are the lowest layer of abstraction in intelligent control. They define how to design intelligent vehicles to work in groups with other intelligent vehicles.

4.2.2.2 NATO STANAG 4586

In 1998, a NATO expert team, composed of members of government and industry (CDL Systems 2010), started working on the development of the standard STANAG 4586 (Compliant Ground Control System for UAV) (Defense Update 2007), which was ratified by NATO in 2002 for the communication and interoperability of its UAV.

The search for interoperability between unmanned systems is essential when meeting objectives in military terms. The line of development should be focused on interoperability between land systems, aerial systems and elements of control, command, communication, computer and intelligence (C4I) (STANAG 2004).

STANAG 4586 was developed as an interface control definition capable of defining a common number of data packets for two new interfaces (CDL Systems 2010):

- A data link interface among ground control stations and aerial vehicles; and
- A command and control interface among ground control stations and C4I systems.

According to Cummings et al. (2006), STANAG 4586 is the only standard that promotes interoperability in control networks of UAVs. There are five interoperability levels defined in this standard (Defense Update 2007):

- Level 1: Reception/transmission of data packets related to UAV.
- Level 2: Received live data about intelligence, surveillance and reconnaissance.
- Level 3: Control and monitoring of data packets of UAVs in addition to the reception of intelligence, surveillance and reconnaissance and other data.
- Level 4: Control and monitoring of UAV, except from launch and recovery.
- Level 5: Control and monitoring of UAV including launch and recovery.

STANAG 4586 supports Electro-Optical/Infrared, Synthetic Aperture Radar, communication transmission and data link interface resources.

4.3 Computer Science Standards

4.3.1 *CORBA*

CORBA is a standard that provides a platform for the development of distributed systems. It allows an easy RMI under an object-oriented paradigm. CORBA is defined by the Object Management Group (OMG), which defines APIs, communication protocols and all necessary items to ensure interoperability between different applications running on different platforms. CORBA uses an IDL to specify the interfaces through their functionality. This is a way to indicate how CORBA data types must be used in implementations of client and server.

All this means that CORBA is a kind of middleware (platform of distributed services, independent of the operating system) that guarantees success in the transit of data across different platforms and applications. It is applied in RTS and is efficient enough for any kind of problem. The main features of this standard are:

- It is a distributed object standard.
- It specifies the architecture the system should have, is flexible and heterogeneous.
- Interoperability.
- Scalability.
- Transparency, facilitating client–object communication (Vinoski 1997).
- Naming service.
- It sets a minimum object model.
- Each object implements an interface.
  - The definition of interfaces is made through the IDL, making it independent of the programming language.
  - The reuse in software is achieved through interface inheritance.
  - Multiple inheritance.
  - The details of an object's implementation cannot be accessed.

4.3.1.1 Components

- The Object Request Broker (ORB) is the CORBA object manager and is part of its core. It allows for the invocation of static and dynamic objects. It can operate without the services and facilities provided by CORBA. It handles the invocation and search for remote objects using dynamic methods for the invocation. It is responsible for giving back the object attributes of the object accessed through the IDL of the object (Vinoski 1997). Locally, it also collects information on the objects to pass to other ORBs and handles local computer security (Fig. 4.2).
- IDL, Language for defining interfaces. Since it is a declarative language and not a programming language, it defines interfaces independent of the implementations of objects.
- Dynamic Invocation Interface (DII). Generic Stub. Client side.

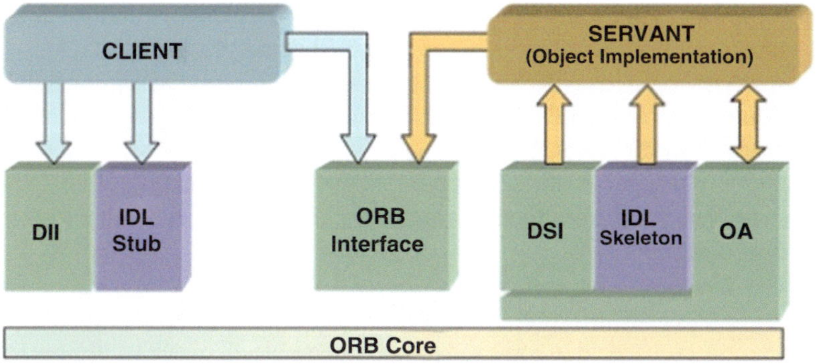

Fig. 4.2 CORBA architecture (University of Seville 2010)

- Dynamic Skeleton Interface (DSI). Generic skeleton. Server Side.
- Both DII and DSI are based on the interface repository, which is a CORBA object that contains information on the object's interfaces and their types. It allows applications to access this information in a static or a dynamic way. The main advantage is the support given to the dynamic calls.
- The implementation repository is required when the objects are persistent. Most general purpose ORBs provide a repository of implementations that supports indirect connections for persistent references. This characteristic solves the problem of direct connections for persistent references. It has also a bad point; it slightly reduces the good working of the first invocation from client to server. It also offers various modes for the automatic activation of server objects (Henning and Vinoski 1999).
- The object adapter is the bridge between the ORB and CORBA object implementations. This allows it to make requests to an object without knowing its interface, since the object adapter adapts the object's interface to that expected from the object making the request.
- Communication protocols between ORBs. CORBA is based on the protocols GIOP (General Inter-ORB Protocol) and the standard protocol IIOP (Internet Inter-ORB Protocol). GIOP specifies the types of messages and the format to transport requests between ORBs. IIOP specifies the way TCP/IP is implemented over GIOP. Thanks to these protocols, ORB can be integrated even if it comes from different developers.

4.3.1.2 Services

There is a large set of standard services offered by CORBA (OMG 1998). These services are added to the ORB interface to complete it; however, they are optional. The most important include:

- **Concurrency Service.** Mediates concurrent access to an object such that the consistency of the object is not compromised when accessed by concurrently executing processes.

- **Event Service.** This defines two roles for objects: the supplier and the consumer. Consumers process information in the events that are produced.
- **Naming Service.** This is the main mechanism for objects that will be invoked by most customers from an ORB-based system.
- **Persistent State Service.** Replaces the persistent object service. These are interfaces that provide persistent information, namely data objects stored in databases.
- **Property Service.** Can attach dynamic properties to objects outside the static IDL-type system.
- **Security Service.** The security service of CORBA provides various security policies to cater for different needs that lead to a secure architecture. CORBA's security can be used in a wide range of systems. It also allows the reuse of its own security protocols. These include:

  - Authentication and identification of objects or users (i.e. verifying that they are who they seem).
  - Access control and authorization.
  - Security audits.
  - Secure communication between objects.
  - Non-repudiation policy

  The CORBA security service is included in the safety process of OMG. Among the OMG security specifications, we can find:

  At an API level:

  - ATLAS (Authorization Token Layer Acquisition Service)
  - RAD (Resource Access Decision Facility)

  In CORBA's infrastructure:

  - CSIv2 (Common Secure Interoperability, version 2)
  - CORBA Security Service

- **Time Service.** Allows an object to ascertain the time along with an estimated error associated to it.
- **Trading Object Service.** Facilitates the search for objects, services, features, functionalities and so on.

4.3.1.3 Application Examples

Some frameworks exploit the features of CORBA for telerobotic systems, whereas some applications may be based on the manipulation of complex systems remotely (Bottazzi et al. 2002).

CORBA is commonly used in telecommunication robots in real time as well as to keep track on them. At the University of Auckland, researchers tested the LEGO Mindstorm and Khepera models to demonstrate the reliability of a design for the distributed control of robots using CORBA (Woo et al. 2003).

The Institute for Computer Design and Fault Tolerance at the University of Karlsruhe in Germany presented a distributed software architecture based on

CORBA for the autonomous service robot Albert2. The development was focused on the modularity and integration of learning aspects (Knoop et al. 2004).

The research group there proposed a system for controlling a humanoid robot based on CORBA. Using this architecture in a distributed environment such as a local network, it is possible that various humanoid robots all over the world can share their own modules via the Internet (Takeda et al. 2001).

CORBA has been used to integrate a distributed system of multiple mobile robots in a simulated environment that offers the possibility of a collaborative control (Zhang et al. 2009).

4.3.2 UPnP

UPnP is a set of protocols (Jeronimo and Weast 2003) or an architecture proposed by Microsoft and promulgated by the UPnP Forum (UPnP Forum 2010). The main goals of UPnP are to simplify the implementation of networks at home and in corporate environments and to connect devices automatically to the network without user intervention. UPnP allows devices to connect perfectly and thereby simplifies network implementation at home (e.g. data exchange, communications and entertainment) and in corporate environments. It provides a distributed and open networking architecture based on already existing protocols and specifications, such as UDP, SSDP, SOAP or XML (Bray et al. 2008). In addition, it is supported by IP as illustrated in Fig. 4.3. Owing to its independence from any particular vendor, operating system and programming language, APIs connected to a network are able to control, negotiate and exchange information and data easily and transparently to the user. UPnP is independent of the physical medium, and it can work over phone lines, power lines, the Ethernet, RF, IrDA and IEEE 1394.

UPnP enhances the concept of a digital home platform in which all household devices should work together. It aims to control each device in the smart home, from consumer electronics to robots, through home appliances using wired or wireless networks. However, up to now, UPnP has not been widely used to manufacture such devices, and it has most commonly been used in simpler systems such as blinds, turning on lights or alarms.

The main feature of this protocol is that there is no need to configure anything when a device is connected to the network. Device services will be automatically available to be used for other entities on the network. This is the main idea in UPnP: each device (a robot, a router, etc.) is available for every entity on a LAN. To offer its services, the device publishes them using a message-passing protocol. UPnP is able to detect when a new device is added to the network. Devices receive an IP address from the network or they assign their own IP (Auto-IP) if a DHCP server does not exist. They then publish this to the network and every device connected to it in order to provide all interesting information such as logic name, developer, model and serial number or the services they offer. This way, the user does not have to worry about complex configurations; he or she just has to add the device to the network.

To understand how UPnP works, we need to describe the components existing on the network and the required stages, including the protocols, to reach interoperability between all UPnP devices.

4.3.2.1 Components

A UPnP network has three main components: devices, services and control points. Components are described below as based on Jeronimo (2004) and the information obtained from Members of the UPnP Forum (2008):

1. Devices
   UPnP devices are logical containers for a service or set of services, and sometimes for other devices (embedded devices). Embedded devices can be discovered and used independent of the main container. Each UPnP device may offer any number of services. By itself, a device just provides a self-description of its information in an XML device description file, and a device's services are those that provide real functionality and execute the actions.
2. Services
   Services provide real functionality and can invoke actions. Each service may contain any number of actions. Each action has a name and an optional set of input and output parameters. A service has an identifier (URI) that uniquely identifies it among all of services. It may keep variables that represent the current state of the service. These state variables may trigger events if they are defined as evented.
3. Control points
   A control point is a network entity that invokes the functionality of a device. It is capable of discovering and controlling other devices. In client/server terms, the control point will be the client and the server role is assumed by the device. Once the device is found, the control point is capable of:
   - Getting the device description and a list of services.
   - Getting the service's descriptions.
   - Invoking actions to control the service.
   - Subscribing to the service. When a service's status changes, the device sends an event to the control points subscribed to the service.

4.3.2.2 Protocols

This section provides a brief description of the UPnP protocols (see Fig. 4.3) used in these networks:

- **TCP/IP**: This is the connection-oriented communication protocol for the Internet and other similar networks. It is based on the idea of an IP address; in other words, it assigns an IP address to each computer or device connected to the network. TCP/IP provides the basis upon which to build a UPnP network.

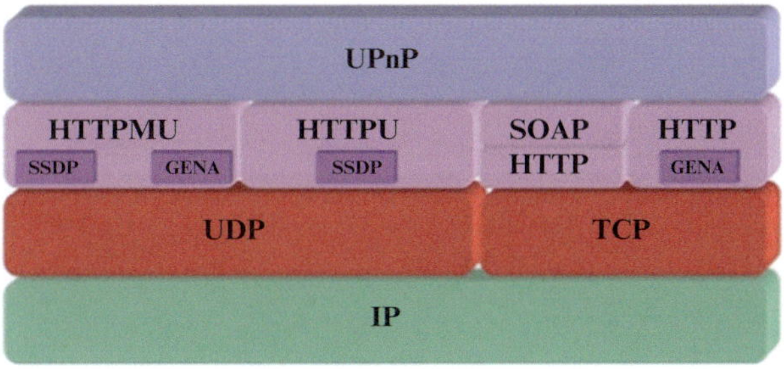

Fig. 4.3 UPnP architecture (de la Pinta et al. 2011)

- **UDP/IP**: This is a connectionless protocol that unlike TCP/IP provides a direct way to send and receive datagrams over an IP network. It supports the HTTPU and HTTPMU protocols described next.
- **HTTP/HTTPU/HTTPMU**: These protocols are essential for building UPnP entities. HTTPU and HTTPMU are the unicast and multicast variants of HTTP. These variants are defined to deliver messages on top of UDP/IP; on the contrary, HTTP works over TCP/IP.
- **SSDP**: This is a protocol that can search for UPnP devices and announce devices and services. Searches and announces used to be made by sending a multicast SSDP message over HTTPMU; however, this may be sent in a unicast message now. When a device receives a search message, it checks the search criteria and if it matches, it will respond with a unicast SSDP message over HTTPU, using the statement "200 OK," which indicates that the request was successful. A SSDP packet is just an HTTP message with the statement "NOTIFY" (to announce) or "M-SEARCH" (to search).
- **SOAP**: This provides a standard mechanism for packaging messages and it defines how two objects in different processes can communicate by exchanging XML files. Each control request is a SOAP message that contains the action invoked and all requested parameters. The reply is another SOAP message that contains the results of the action or the errors as appropriate.
- **GENA**: This defines an HTTP notification architecture that allows transfer notifications between HTTP resources.
- **XML**: This organizes, stores and exchanges information, and its main function is to describe data. It is used in UPnP for device and service descriptions, control messages and events.
- **HTML**: This is a markup language that uses a set of markup symbols or codes to structure text and multimedia documents and to set up hypertext links between documents.

4.3.2.3 UPnP Operation

To describe the way that the protocol operates, we need to show the six basic steps in a UPnP network: Addressing, Discovery, Description, Control, Eventing and Presentation. Addressing may be considered step zero of UPnP networking.

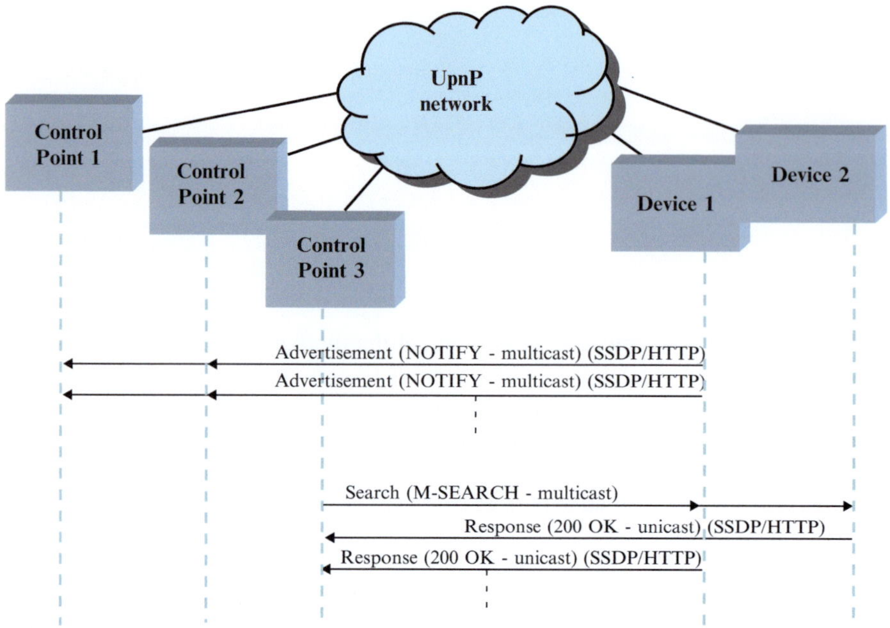

Fig. 4.4 Discovery (University of Seville 2010)

This book presents a simplified version of how UPnP operates. However, these steps are detailed in the UPnP Device Architecture document (Members of the UPnP Forum 2008):

1. Addressing
   Devices and control points must obtain an IP address before they can join to a UPnP network; therefore, when they are first connected to the network they must search for a DHCP server to get an IP address or use Auto-IP to obtain an address. UPnP entities may retrieve an IP address from a DHCP server; to that effect, both devices and control points must have a DHCP client. If the network does not have a DHCP server, devices and control points must use Auto-IP to get the IP address. Through this mechanism, the device takes a random address in a range established by the ICANN/IANA. Once the address has been allocated, the entity checks it using the ARP protocol, and if it is being used on the network the device will get another IP address.

2. Discovery
   This step defines how a device announces its presence and how a control point discovers devices using the SSDP (Fig. 4.4). The Discovery stage allows control points to find devices and services and to obtain information about them.

   - Advertisement. Once devices are added to the network, they multicast messages to announce their embedded devices and services to control points through NOTIFY packets. These messages do not require a reply and are

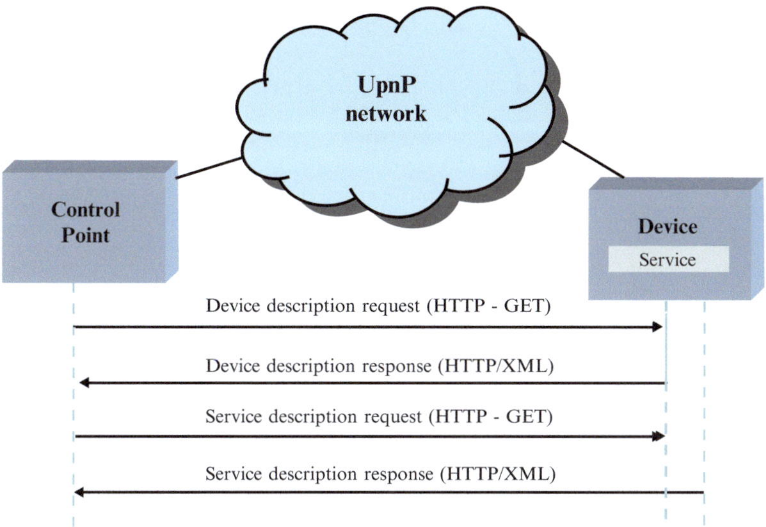

Fig. 4.5 Description (University of Seville 2010)

resent periodically when devices renew their advertisements. Through these messages, control points may retrieve the descriptions devices and then may control devices and retrieve the descriptions of services to manage these services, invoking actions and subscribing to events.

- Search. This procedure allows control points to search for devices on the network. Control points may search for specific devices or services through M-SEARCH messages. Responses from devices are needed, and these contain discovery messages similar to the advertisement ones; however, the responses are unicast because devices know the control point address.

3. Description
After the Discovery step, the control point retrieves the information from the discovery message, i.e., a universally unique identifier and a URL of the device's UPnP description. The Description step consists of retrieving the description of the device and its capabilities (service description) from this URL. The descriptions of the devices and their services are stored in XML documents. A device description contains device information, a list of the services provided by the device and a list of their embedded devices. A service description includes detailed information about the device's service, the actions provided by the service, as well as input parameters and output state variables. To get the description files (see Fig. 4.5), a control point sends an HTTP request using the GET method to the URL contained in the discovery message that had previously been received by the device. When it receives the request, it replies with an HTTP message that contains the device's description in the message's body.

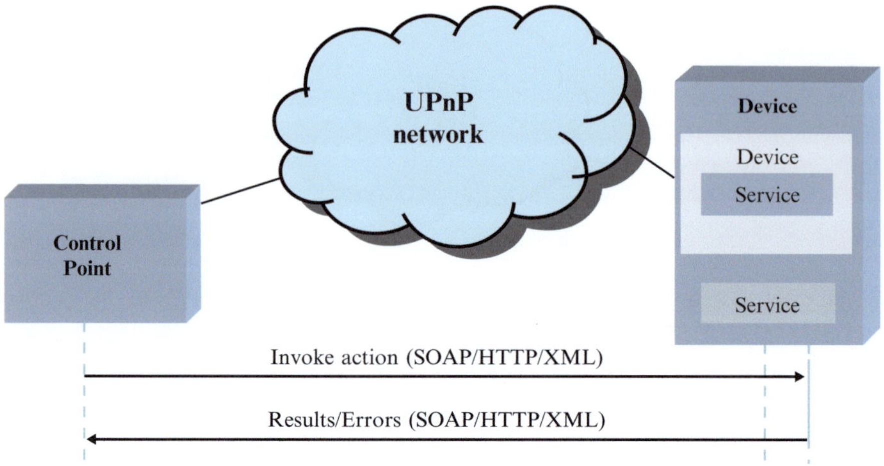

Fig. 4.6 Control (University of Seville 2010)

4. Control
   This is the step in which the control points invoke actions on the devices' services. Once a control point has all the information about a device and its services through their descriptions, it will be able to control this device by invoking actions. The Control step is based on the SOAP, which uses XML and HTTP to provide web messaging and RPC. To invoke a specific action, the control point must send a SOAP request using the POST method to the device's service. Then, the device will respond with the results or the errors obtained as a consequence of the invocation. This stage is illustrated in Fig. 4.6.

5. Eventing
   Eventing can notify a control point when the state of a device changes. As explained above, a service description contains a list of variables that models the state of the service. If any of these variables is configured to report an event (evented variable), the service publishes updates when any of these variables are modified.

   Eventing uses a publisher/subscriber model in which the control points can subscribe to events sent by a service. The services publish event notifications to subscribers. An event is a message sent from a service to the subscribed control points. The events inform the subscribed control points about the state changes in the services.

   A control point that wants to be notified about the changes in the variable's state subscribes to an event source by sending a subscription request to the URL of the events, which is contained in the corresponding device description. If a service accepts the subscription request, it responds with a SID and the duration of the subscription. The SID allows the control point to refer to the subscription in subsequent requests to the service, such as renewing or cancelling the subscription (Jeronimo 2004). Eventing protocol is a GENA and is used over the TCP layer, which guarantees message delivery to the subscriber. Figure 4.7 presents a diagram of this process.

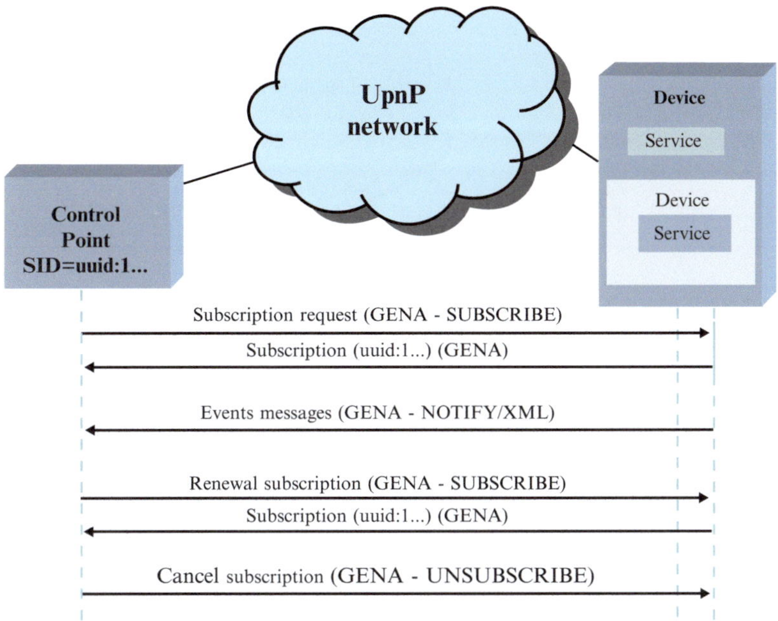

Fig. 4.7 Eventing (University of Seville 2010)

6. Presentation
   Presentation is considered as an optional step. A control point may monitor a device or check its status through the presentation of a webpage in HTML. If a device has a presentation page, control points may load presentation pages in a browser and these allow users to check and control the device. To retrieve a presentation page, the control point issues an HTTP GET request to the presentation URL and the device returns a presentation page (Microsoft) (see Fig. 4.8).

   It is also interesting to review UPnP applications developed in recent years to understand the interoperability provided by this architecture. For example, Maestre and Camacho (2009) state that different consumer electronic devices have been developed using UPnP architecture. De la Pinta et al. (2011) show that the Roomba robot has been successfully integrated into a UPnP framework. In addition, UPnP AV devices have been integrated into an OSGi platform (Kang et al. 2005). Another example of UPnP interoperability is the success of the DLNA protocol in multimedia services, which is derived from the UPnP architecture.

4.3.3 Jini

Jini is a service-oriented architecture developed by Sun Microsystems that provides an infrastructure for defining, publishing and searching for services on a network. Service Discovery (similar to UPnP service) is the main feature in the

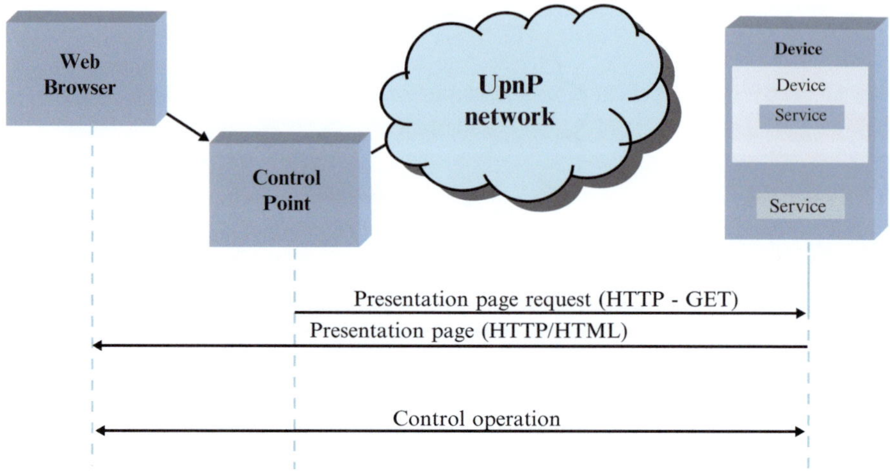

Fig. 4.8 Presentation (University of Seville 2010)

Fig. 4.9 Jini architecture diagram (University of Seville 2010)

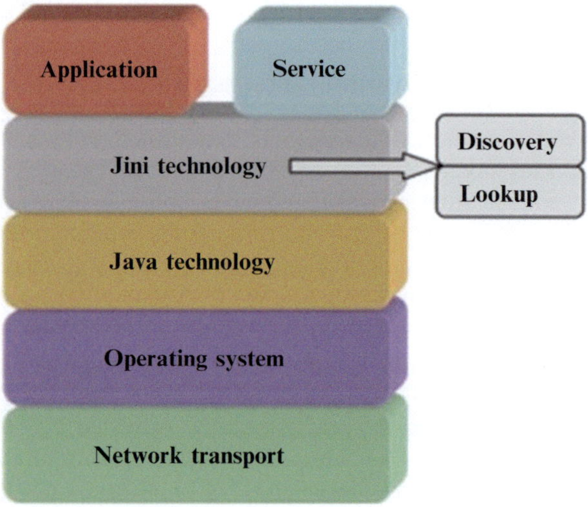

Jini technology, both in multicast mode and search mode for specific services. Jini uses the multiplatform feature from the Java platform to provide universal services, and it registers each one of them as serialized objects (service proxy) with its own interfaces. A Jini architecture diagram is shown in Fig. 4.9.

The main aims of this platform are discussed in Arnold (1999), which exposes its immediate services availability, the hardware abstraction, the service-based architecture and the simplicity. Jini is an easy protocol (Morgan 2000) as explained in Fig. 4.10.

- When a device is connected, it looks for a lookup service (Discovery) with which to register.

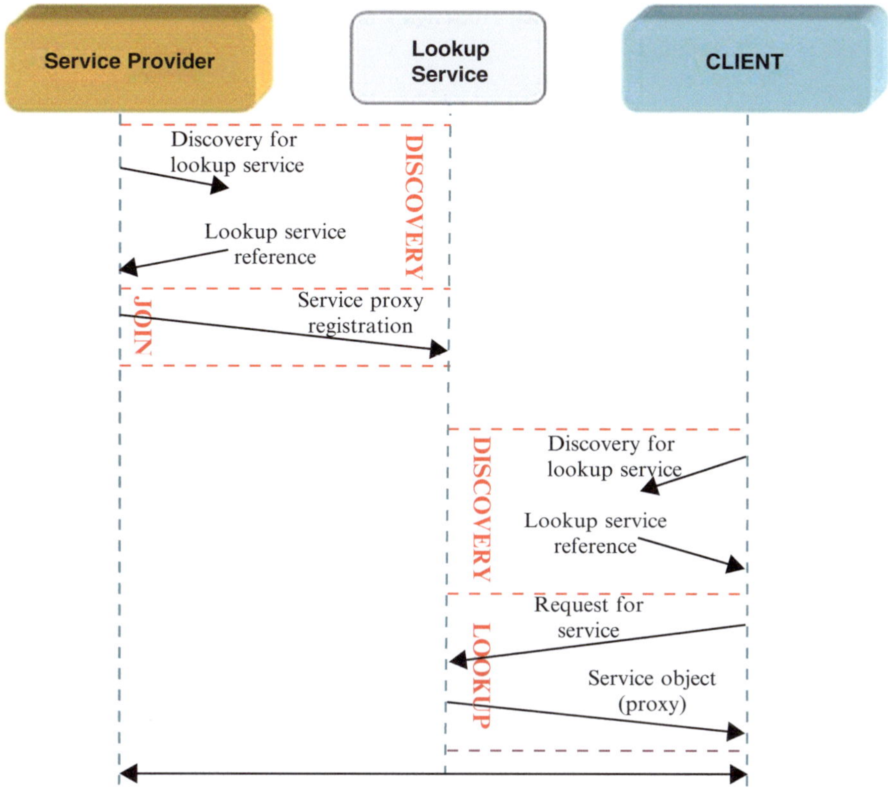

Fig. 4.10 Jini events (University of Seville 2010)

- When a service provider locates a lookup service, it joins to it (Join). The service uploads a service proxy that a client would need to use its services, and the lookup service stores it.
- When a client needs to locate and invoke a service, it asks the service for the lookup service, and it gives back the service proxy mentioned above.
- Then, the client is able to interact with the service provider (during an specific time, in a shared way or in a exclusive one) through the proxy.

The purpose of the Jini architecture is to organize devices and software into groups inside a distributed and dynamic system. This simplifies the access, management and maintenance of each service offered by service providers. Some interesting concepts in a Jini system are presented below:

1. Services
   A Jini system consists of a set of services that can be used to perform a particular task. A service is an entity that can be used by one person, one program or another service. It may be a calculation, saved data, a communication channel with another user, a software filter, a hardware device or another user.

Services communicate with each other using a service protocol (set of interfaces written in Java language).

2. Lookup Service
   Services are found through a lookup service. This is the central mechanism for the system and provides a mapping service that indicates the functionality provided by the services. A service is added to a lookup service using the discovery and join protocols. The service locates an appropriate lookup service (using the discovery protocol) and then joins to it (using the join protocol).
3. Java RMI
   This is a mechanism provided by Java to invoke remote methods. RMI is a Java extension of RPC. It provides remote communication between programs written in the Java programming language. The RMI subsystem also implements reference counting-based distributed garbage collection to provide memory management facilities for remote server objects.
   RMI allows not only data to pass from one object to another through the network, but also whole objects to be sent and received, including their codes. Much of the simplicity of the Jini system is because of this ability to move code through the network, encapsulated in an object.
4. Security
   The Jini security model is based on the concepts of a master list and an access control list. Jini services are accessed by an entity – the principal – that generally refers to a particular user in the system. The access of an object to a service depends on the contents of the access control list associated with the object.
5. Leasing
   A lease grants access to a service for a certain period of time. Each lease contract is negotiated between the service user and provider as part of the protocol service, and it is released if the contract is not renewed.
6. Transactions
   A transaction can group a set of atomic distributed operations into a single unit. If one or more operations fail, the transaction is aborted and no partial results are written.
7. Events
   Jini supports distributed events. Objects may register to events in other objects. When an event occurs, a notification is sent to the objects that have been registered.

4.3.4 Web Services (WS)

WS is a technology that allows websites to use distributed applications and offers features such as access to the information and functionalities of any platform. At first, they were created to meet the need to standardize communication between different platforms and programming languages because earlier attempts such as CORBA had little success. In the case of CORBA, this was because there are certain limitations for more complex applications that require a security control or transaction management.

WS provide a standard means of interoperating between different software applications, running on a variety of platforms and frameworks. WS are functions or procedures that can be accessed via the web. Regardless of the programming language of the service and its platform, they enable the exchange of data and provide services between different applications.

Such a degree of interoperability is only possible using open protocols. WS are mainly used with HTTP because this is widely used and is rarely blocked by firewalls. WS are a set of protocols and standards used to exchange data between applications, and they are used on important websites for tasks such as e-commerce, web browsers and computer services by companies such as Google, eBay or Amazon. The W3C is responsible for managing the specifications. The main features of WS technology and its advantages and disadvantages are listed below:

- It is supported by any platform and any programming language.
- It is a W3C standard.
- It provides functionality to websites.
- It uses HTTP to transport data.
- It uses standard elements for each of its components (SOAP, UDDI, Web Services Definition Language (WSDL) and XML).
  - One of the main advantages of WS is that they allow applications to communicate efficiently, regardless of the platforms used, offering greater interoperability. WS use standards and text-based protocols, which allows a better understanding and easier access to the data exchanged. They also use HTTP to allow the information to pass through firewalls without major complications. This fact together with the use of XML promotes interoperability.
  - However, WS are much less efficient than are CORBA or RMI because they make use of formats based on text, such as XML, which are not the best options to process tasks. Nevertheless, new WS standards may define more optimized protocols. Also they are not as developed as standards such as CORBA. Both HTTP and XML have a high run-time cost compared with other distributed applications approaches. Skipping the firewall security can also be seen as a drawback.

4.3.4.1 Components

WS use text-based standards and protocols, and this involves the components listed below. Figure 4.11 shows the diagram of the interactions between the entities and flows of the incoming and outgoing data of each component.

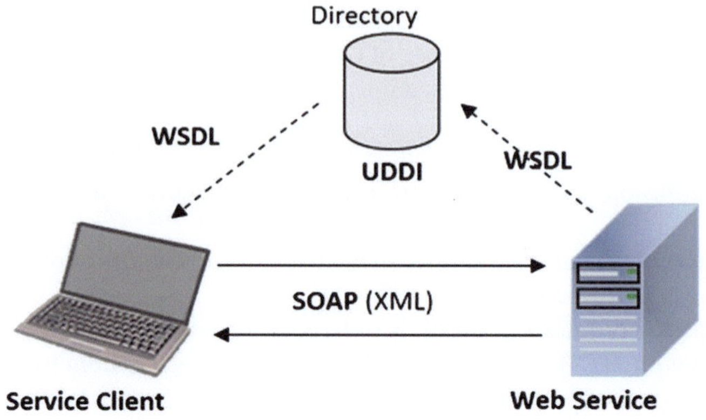

Fig. 4.11 WS communication architecture (University of Seville 2010)

1. WSDL
   It is desirable that WS have information on the operations and data types involved. For this reason, WSDL is used. This is a standard adopted by the W3C that defines the public interface of WS. It is structured as follows:

   - Ports (<portType>): these describe the operations provided by WS. Its function is similar to an object-oriented class.
   - Messages (<message>): these define the data involved in an operation, where each message can have one or more parts. It is considered one of the parameters used in object-oriented programming.
   - Types (<types>): these define the data types involved in WS, using XML Schema, an XML language that accurately describes the structures and constraints of the XML file. It has been in the W3C since 2001.
   - Links (<binding>): these describe the message formats and the protocols for each one of the ports.
   - Operations (<operations>): these can be one-way, request-response (makes a request and waits for a response), request-response (receives a request and makes a response) or notice.
   - Services: these define a set of web service ports.

2. UDDI
   To register and publish WSDL we use Universal Description, Discovery and Integration (UDDI). This is a standard developed for the publication and registration of WS. Its way of working is similar to a database and has two different parts:

   - Registration of business:

     - White Pages (Overview)
     - Yellow Pages (categories of services)
     - Green Pages (business rules)

   - Registration of services

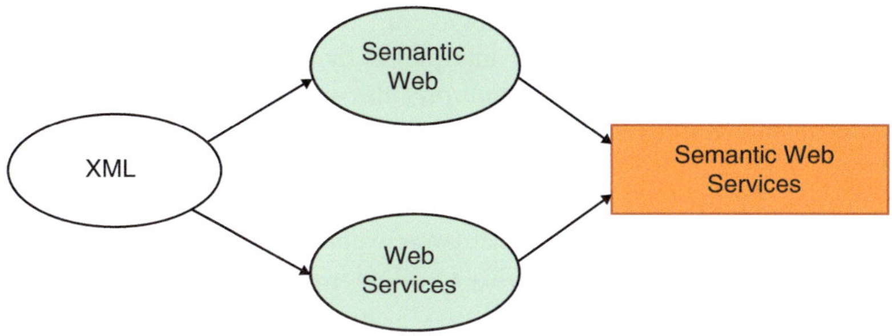

Fig. 4.12 The emergence of SWS (University of Seville 2010)

3. SOAP

   In addition, there was a need to define the way of exchanging data between different processes on different machines. For this task, we use the SOAP, which defines the format of the messages to send. It is independent of the transport protocol. The elements of a SOAP message are (Daconta et al. 2003):

   - Encapsulation of the message.
   - Description of the data coding.
   - Body, which contains the specific message of the application.

4.3.4.2 Applications

Websites ask WS for a series of functions. They are currently used in almost all websites and they provide most logic to the website. Another possible application of WS is for the control of robots. WS are used to control robots from anywhere in the world via the Internet through a user interface, which will provide the services offered by the robot as well as its status (Levine and Vickers 2001).

4.3.5 Semantic Web Services (SWS)

SWS were derived from the combination of WS with the emergence of the semantic web (Fig. 4.12). Tim Berners-Lee created the semantic web states that the "*Semantic Web is not a separate web but an extension of the current one, in which information is given well-defined meaning, better enabling computers and people to work in cooperation*" (Berners-Lee et al. 2001). WS meet the requirement of a specified syntax; however, they have a lack of semantics so they cannot resolve ambiguities. This is solved by using SWS, optimizing this way the reuse of WS and creating smarter websites, resulting in the concept of Web 3.0. This simplifies the sharing and integration of web resources.

To represent knowledge, ontologies that structure information, resources or services based on the meaning of words emerge. This allows computers to interpret and process this information to work automatically.

The languages of high-level ontologies are backed by a formal logic, which makes sure that the ontology can be interpreted by the machines. This means that the computer and its software can interpret the semantics of the model without direct human intervention. The ontological software rises to the level of human conceptual knowledge; humans do not have to descend to the machine's levels (Daconta et al. 2003).

SWS are an important line of the semantic web, which aim to describe not only information but also WS's functionality ontologies and procedures: its inputs, outputs, conditions for implementation, effects produced or steps followed. These machine-processable descriptions will automate the discovery, composition and implementation of services, as well as the communication among them. The semantic web has emerged to provide the syntactic web with semantic intelligence and has the following main features:

- Automatic data interpretation.
- Ontologies as data models.
- Discovery, selection and automatic service composition.
- Service implementation through the web.

4.3.5.1 Required Functionalities

- Publication of service descriptions.
- Services discovery.
- Service selection.
- Composition of services.
- Resolution of problems caused.
- Implementation of automated services.
- Monitoring of implementation.
- Compensation.
- Substitution of services for similar ones.
- Verification of implementation.

4.3.5.2 Main Technologies

- Web Ontology Language (OWL-S). This is an ontology based on OWL, which is a markup language for publishing and sharing data using ontologies. It was created by DARPA (2007), which is part of the US Department of Defense, where they automate tasks such as the discovery, invocation and composition of WS.
- Web Service Modeling Ontology (WSMO). This is a conceptual model for the relevant aspects of SWS and it belongs to the European Semantic Systems Initiative. The WSMO working group includes the technology of Web Service

Modeling Language, which formalizes the WS that model the ontology (Lara et al. 2004). Its main components are:

- Goals. These are the customer's aims when they access the web service.
- Ontologies. A formal description of the semantics used by all components.
- Mediator. These are connectors that provide interoperability among different ontologies.
- WS. These can include the functional and usage descriptions of WS.
- OWL-S has a weak point in the architecture because it is undefined. It also has little development in comparison with WSMO. Its difficulty is also higher and less intuitive than WSMO is. However, its groundings of use are well developed. However, WSMO is not mature in key areas of use. It has a robust and flexible architecture for the consumer in contrast to OWL-S. It has defined important aspects such as languages and mediation. There are also plans to automate the creation of WS based on WSMO to semi-automate this process, thereby saving money, time and resources; the same as in the IRS III project.

4.3.6 RMI

RMI emerged from the need to communicate among different objects, and it is implemented on different machines as happens on distributed systems. Therefore, this technology is a remote invocation of Java objects. The initial version of Java RMI required a JVM in both the origin and destination machines (Cheng-Wei et al. 2004).

After the RMI-IIOP was developed, it was added to the RMI, providing it with the best features of CORBA. RMI is pure Java and since it does not support other languages, CORBA emerged. The adaptation to a distributed system has not prevented the continued development of RMI as a secure system. The main characteristics of RMI are:

- Simple, easy to write and easy to maintain.
- Transparency, because the distribution of objects and parameters passing is transparent to the programmer.
- Pass an object by value (as parameters of methods).
- The definition of interfaces is done directly in Java.
- Implementation in Java.
- Independence of the communication protocol.
- Separation between interface–client and implementation–server.
- Naming service.

4.3.6.1 Architecture

RMI is a layer architecture made of a stub/skeleton layer, a remote reference layer and a transport layer. The programmer only interacts with the application layer. The

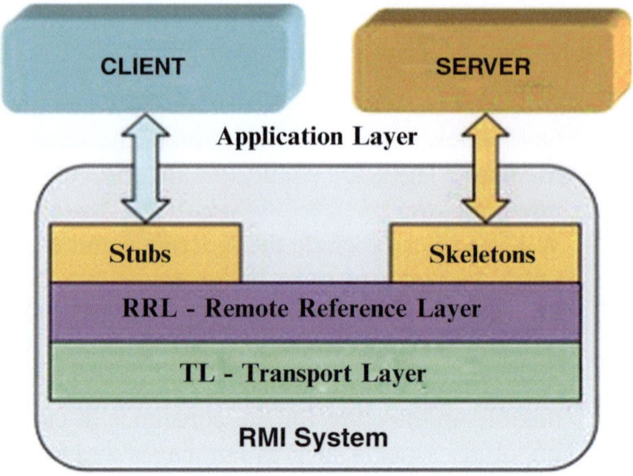

Fig. 4.13 RMI architecture (University of Seville 2010)

RMI system manages the three previous layers (see Fig. 4.13), which could be replaced by others with the same function without altering the rest.

4.3.6.2 Components

1. IIOP

   RMI allows the programming of CORBA servers and applications via the RMI API. It is possible to work entirely in the Java programming language using the Java Remote Method Protocol as a transport or to work with any other CORBA implementation using IIOP Java RMI over IIOP.

   RMI-IIOP is designed for developers who program in Java and want to use the RMI interfaces using IIOP as the transport layer. The RMI-IIOP interoperability with CORBA objects implemented in other languages is available only if all the remote interfaces have been previously defined as Java RMI interfaces (Oracle 2010).

4.3.6.3 Application

At the University of Bielefeld, Germany, one research group has integrated memory-based software for the development of autonomous robots. This is an approach to an architecture of autonomous mobile robots operating in human environments. It replaced the use of data on a closed chain based on the long- and short-term memory. RMI was used for the exchange of critical information, such as the module that controls the hardware. RMI also allows the system to estimate when the configuration has been completed. The system can then send information on the result of the configuration (Spexard et al. 2008).

Westhoff et al. (2004) focuses on task-level programming and monitoring robots in their daily operations. It is not a framework limited to robots and it could be used in other distributed environments. During its development, the authors took advantage of technologies available in Java, such as Jini, RMI and Java Native Interface.

Woo et al. (2003) supported Java RMI over Bluetooth, GPRS and WLAN technologies. As a conclusion of this, the good work of Java RMI was tested in heterogeneous wireless environments, allowing parallel and distributed control.

In a study by researchers at the Information and Communications University in Korea, RMI is used to access external ontologies in the development of a self-expandable software. This kind of software is useful for intelligent robots for two reasons. First, they study their environments and then they decide their appropriate behavior based on what they have learnt about their surroundings.

DEVS/RMI is a distributed, self-adaptive and reconfigurable simulation environment for engineering studies. It is based on the standard implementation of DEVS, in which Java RMI supports the synchronization of local and remote objects. It is designed for the intensive testing of programs, and this is the reason for it supporting dynamic models (Zhang et al. 2005).

4.3.7 Other Computer Science Standards

4.3.7.1 DH Compliant

DH Compliant (DH Compliant 2010) is a system providing interoperability between all devices existing in a home network. It is based on the UPnP architecture and is currently under development by the University of Oviedo, the University of Seville and a consortium of companies composed of Ingenium (Ingenium 2010), Domotica Davinci (Domotica Davinci 2010), MoviRobotics (MoviRobotics 2010), (Applied Research Associates) (ARA 2010) and the Cartif Foundation (Cartif 2010). The main goal of DH Compliant architecture (Fig. 4.14) is to integrate consumer electronics devices, robots, sensors and other interesting components that may be useful in a home automation framework.

The aim of the DH Compliant system is development and implementation that allows the integration of service robots within the digital home. This architecture will provide interactions between robots and digital homes to make life easier, more secure and more comfortable. This protocol integrates the intelligence of a UPnP control point and the functionality of a UPnP device in a single DHC device. This entity network is managed by other entities that provide new services such as the localization service, energy-saving service and the service for collaborative tasks between robots.

4.3.7.2 OSGi

OSGi (OSGi Alliance 2003) is a module system for the Java environment that implements a components model, which needs JVMs. OSGi is based on a layer model that includes, among others, a bundles layer that provides the applications and components as packages (i.e. jar files), a services layer that provides communication between bundles through Java objects, and modules and security layers.

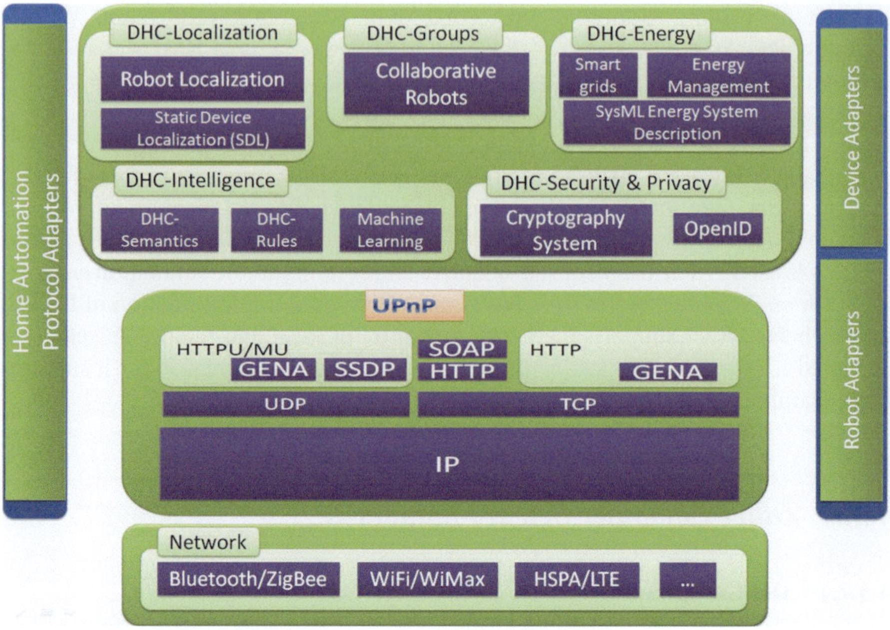

Fig. 4.14 DH Compliant architecture (University of Oviedo 2010)

OSGi may be a good alternative for the development of complex systems because of its versatility and cross-platform feature (one JVM in each network node would be necessary to run the application). Any framework that implements the OSGi standard must provide applications modularity to decompose the application into small packages. Each package is a collection class (jar and settings files). The framework is conceptually divided into the following areas:

- **Bundles.** This is a set of Java classes and additional resources.
- **Services.** This connects bundles dynamically. There is also an API for services management.
- **Lifecycle.** This is the API to manage the lifecycle and it spans install, start, stop, update and uninstall.
- **Modules.** This defines how bundles import and export code.
- **Security.** This limits bundles' functionality to predefined capabilities.
- **Execution environment.** This defines what methods and classes are available on a specific platform.

Some examples of OSGi uses can be found in the literature. Gu et al. (2004) discussed an intelligent system (SOCAM) based on ontologies integrated with OSGi to build a system that can deliver and manage context-aware services in a smart-home environment. Meanwhile, Kang et al. (2005) fuse UPnP AV, which is used to provide media services, with OSGi, which manages each UPnP entity as a bundle.

Chapter 5
Robotics Perspective

Alberto Alonso Fernández, Pablo Fernández de Dios, and Jorge Moreno Sánchez

Abstract The service robots market perspective indicates that the incursion of these elements into the home will continue to rise according to a series of factors. On the one hand, these factors include an increase in home automated facilities in new built housing, the largest number of Internet connections in homes and the price decrease of the components that shape the digital home. On the other hand, the service robots forecast should take into account aspects that influence adjacent markets such as the economic crisis in the construction industry. In addition, demographic factors, the lack of skilled workers and rising life expectancy are aspects to bear in mind in the forecast. Therefore, the scientific community must propose a roadmap to settle the technological challenges that are appearing in an unstructured and dynamic environment such as the home.

5.1 Introduction

The price reduction of robotics should allow, as has happened in the world of computing, users to have robots in their homes. According to the Strategic Research Agenda for robotics in Europe (Bischoff and Guhl 2009), the robot population in the world will reach 18 million during 2011. Further, the World Robotics 2010

A.A. Fernández (✉) • P.F. de Dios
Infobotica Research Group, University of Oviedo, Oviedo, Spain
e-mail: alonsoalberto@uniovi.es; fernandezdpablo@uniovi.es

J.M. Sánchez
Domótica daVinci S.L., Tiñana-Siero, Asturias, Spain
e-mail: jorge.moreno@domoticadavinci.com

I.G. Alonso et al., *Service Robotics within the Digital Home*, Intelligent Systems, Control and Automation: Science and Engineering 53, DOI 10.1007/978-94-007-1491-5_5,

© Springer Science+Business Media B.V. 2011

(IFR – Statistical Department 2010) report indicates that the forecast for 2010–2013 is 80,000 new service robots for professional use and some 11.4 million service robots for personal use.

The application of service robots for personal and private use will be focused on the accomplishment of domestic or social tasks. These objectives may lead to a number of ethical, legal or social issues that should be addressed over time in legislative actions and social interactions that support developing new market areas.

5.2 Ethical, Legal and Societal Issues

Service robots can be defined as robots that operate semi or fully autonomously to perform services useful to the welfare of human beings and work equipment, excluding manufacturing work. The aid offered by these robots to people in their daily work, in their domestic tasks or as part of assistance to the handicapped and the elderly raises questions about the uses of spaces in which humans exist and in their collaboration with humans.

5.2.1 Ethical Issues

The utilization of robots must be focused on helping humans; robots should never replace humans in habitable environments. Likewise, a robot cannot be used to violate intimacy. The top position in any hierarchy of control must be ensured to be human. However, educational robots should not imitate human forms or behaviors in order not to substitute for teachers. Further ethical issues can be derived from the Universal Declaration of Human Rights.

5.2.2 Legal Issues

In this regard, a robot that has stored or transmitted personal information must be subject to legislation that will prevent intrusions into the privacy of users. This can raise other legal debates such as the consideration of the robot as a subject or as private property. Is the robot (or its manufacturer) responsible for its acts? Is the robot's owner the legal person in charge? The responses to these questions remain ambiguous.

In this context, when a robot takes wrong decisions, is the designer, producer, commissioner or user responsible for the inappropriate actions of the robot? The robot's learning process must be controlled by the person who assumes legal responsibility for it.

5.2.3 *Societal Issues*

Some work profiles describe actions that can be efficiently carried out by robots, such as digging or transporting dirt, or performing highly repetitive tasks. These can be an advantage in improvement of quality of such tasks, but can lead to an increase in the index of labor unemployment and thus social dissatisfaction with robots. Finally, the representation of human forms in the body of the robot can also present social rejection or acceptance according to the task in hand.

5.3 Principal Markets

A roadmap is required to ensure a diversity of applications and solutions that incorporate full-scale general autonomous functionality. The Roadmap for US Robotics (Collaborative 2009) considers a period of 10–15 years to see this wide variety of applications.

5.3.1 *Demographic Factors*

Several key factors can help identify trends in order to market service robots. One key factor is the aging population, which can influence the growth of service robotics in both a professional and domestic environment. The aging population will produce a deficit in the workforce, which could then be replaced by robots; however, an increase in the population with mobility difficulties will favor an expansion of the number of domestic service robots in charge of caring for people with limited mobility (senior citizens or disabled persons).

According to a release of the US Bureau of Labor Statistics (2010), and as pointed out by the Roadmap for US Robotics, the number of retired workers as a percentage of the current workforce will double within 20 years. It will pass a ratio of two retirees for every ten workers in 2009 to almost four retirees for every ten workers in 2030.

Similar estimates have been made by the World Bank (Australian Government – The Treasury 2002) for the US, placing it in the middle of an aging workforce (Fig. 5.1). In Europe and Japan, aging is more pronounced (Fig. 5.2), whereas in East Asia and the rest of the developing countries ratios will remain lower than they are in developed countries (Fig. 5.3), although the increase in the ratios for most are similar to those for the US, the UK and France.

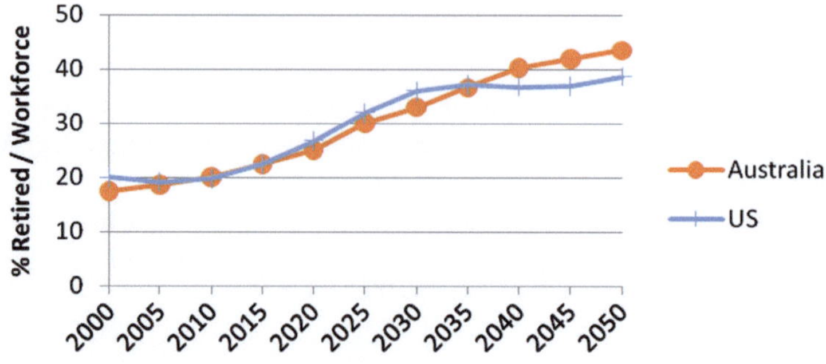

Fig. 5.1 Ratio of retirees to workers (medium level)

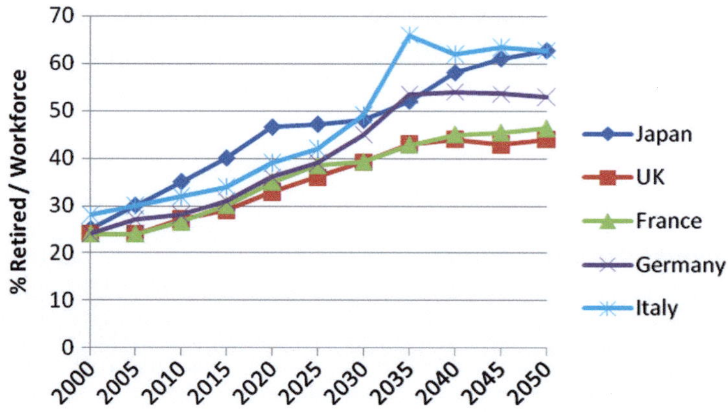

Fig. 5.2 Ratio of retirees to workers (high level)

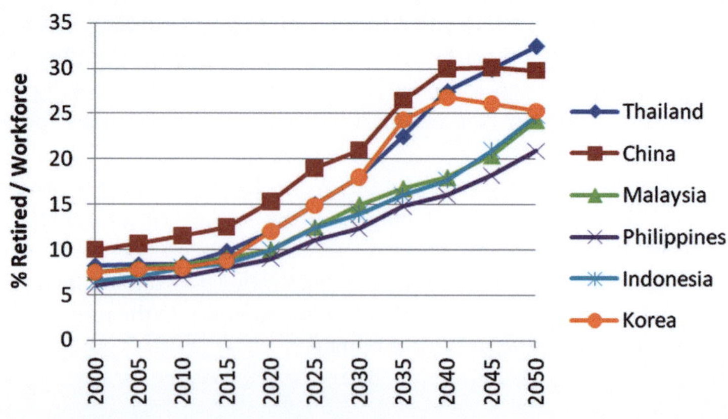

Fig. 5.3 Ratio of retirees to workers (low level)

5.3.2 *Market Pull*

The professional and domestic service robots sector will be present in different markets and can be identified through existing projects, demands and market requirements.

- **Healthcare and Quality of Life**. Current robotic applications provide teleoperated solutions. The Intuitive Surgical's daVinci surgical system (Ballantyne 2002) represents an advance in laparoscopic surgery, but it actually conceals the vast potential of robotics in the healthcare market to help control costs and empower healthcare workers. Service robotics enable aging citizens to live longer in their homes by facilitating household chores or removing architectural barriers.
- **Manufacturing and Logistics**. Service robots can perform logistical tasks in a wide variety of environments such as hospitals, factory warehouses and transport networks. Simple robots such as transit trains for passengers at airports are only a reflection of the robotics applications in small-scale (public transport in hospitals, offices and places) and large-scale logistics (autonomous road transport of goods and people).
- **Automotive and Transportation**. Progress in the automotive and construction of autonomous vehicles has begun to take its first steps with vehicles at Segway (Segway 2010) or Honda U3-X (Honda Motor 2010) set to be introduced to private transport. These vehicles use the movement of passengers for navigation but they are not autonomous. Vehicle manufacturers have created intelligent cars capable of helping drivers to park, counter-steer or improve their energy efficiency. Initiatives such as the DARPA Urban Grand Challenge (DARPA 2007) in the US, the ARGO research project (Parma University 2002) in Italy or the EUREKA Prometheus Project (Williams 2002) invest large financial sums in the development of autonomous vehicles. Its aim is to reduce the number of accidents caused by human error and reduce fuel consumption.
- **Homeland Security and Infrastructure Protection**. Surveillance robots can play a role in border protection, search and rescue, port inspection and private security.
- **Entertainment and Education**. In this area, the imagination is the limit. In 20 years, the utilization of motion simulators, rollercoaster educational aids, personal sports trainers and novel games is forecast.
- **Energy and Environment**. Increasing productivity while reducing costs is the common task of service robotics, and task automation contributes to reducing energy costs and monitoring energy expenditure.

5.4 Scientific and Technical Challenges

For robots to reach the desired levels of development and to comply with the requirements of tasks they can do, a series of technological challenges must be met. These technological challenges are focused on reducing time and energy costs, increasing the quality of products manufactured by robots and improving human–robot interactions.

5.4.1 Positioning

This process is relevant to robot motion in a defined place. The land on which robots must move vary: ground, water, air and space. Current position is based on the environment and 2D models. 3D navigation is one of the most representative challenges. 3D models not only must contain world geometric layers but also maps containing semantic information about the task as well as objects and features of the environment in which it is to be performed. These new navigation and positioning systems must include object affordances because achieving semantic 3D navigation will require novel methods for sensing, perception, mapping, localization, object recognition, affordance recognition and planning.

5.4.2 Manipulation and Grasping

Manipulation and grasping are required abilities in order to handle physical objects. Such applications require a robot to interact physically with its environment. Nowadays, autonomous manipulation robots work well in highly controlled industrial environments but cannot handle the environmental variability and uncertainty associated with dynamic and unstructured environments. To introduce robots into these changing environments, it is necessary to ensure that no catastrophic failure can occur; this implies the need for the robot to perceive its unstructured environment in order to accommodate changes in planning and prioritizing tasks dynamically.

5.4.3 Sustainability

The manufacture and deployment of robots must take into account the growing concerns about their environmental and social impact. The physical makeup of robots must adhere to environmental requirements on sustainability with respect to their manufacture, utilization and ultimate recycling. Robot design must include software and other aspects that ensure the minimum consumption of resources during their lifecycles.

5.4.4 Autonomy

Autonomy can be defined as a system's ability to independently perform a task, carry out a defined process, or make a system adjustment. Autonomous robots must be pre-programmed, but in most domains autonomy must be limited to ensure the hierarchy of control for humans.

5.4.5 *Configuration*

A robot's configuration is designed to specify a task or set up a system; a change in a robot's setup is usually performed by an operator when the system is not in operational mode. It is done mainly through programming, instructing, initializing or testing. The technology challenge is to minimize the need for manual configuration. This process will be simplified with new human–robot and robot–robot interfaces.

5.4.6 *Adaptation*

Robots must be able to make some changes in implementation or planning by themselves, without human intervention. These changes may take place in the short- or long-term and affect different levels of the system. This adaptation process might involve technological challenges such as cognitive decision making, changing the operational parameters of the software or adjusting hardware depending on the environment.

5.4.7 *Human–Robot Interaction*

Mutual communication between robots and humans should be resolved by common cognitive contexts and visions. This may involve changes in the environmental and physical interactions between robots and people until it reaches a natural interaction between them.

To improve the quality of work and reduce energy costs and time, the cooperation of multiple robotic systems with a common goal must imply complete interoperability between robots. They can interact directly or through modification of the environment. Currently, the cooperative tasks may be predefined or prescripted under a centralized control system. Robots with manipulators will jointly carry out a process in close proximity.

5.4.8 *Dependability*

Dependability refers to the integrity of the robot while it performs tasks safely and reliably. The robot is dependable if it is maintainable, available, robust and secure. The robustness of the materials that compose robots and future self-diagnosis and control applications will be crucial to prevent system degradation and extend the time between maintenance.

5.4.9 Physical Properties

The hardware and software design of a robot depends on the requirements of the tasks and work environments. For example, robots operating in domestic environments must be tailored to a working environment designed for humans. The standardization and modularity of components will increase and design tools will be improved.

5.4.10 Process Quality

This factor determines the quality of the product or task fulfillment as well as the consistency and success level of the robot. Depending on the atomicity of the task, this term can be referred to as the level of task fulfillment. The production of robotic systems will be significantly higher than that of humans in all tasks.

5.4.11 Standardization

To ensure the quality of robotic systems and lower manufacturing costs, the standardization of robotic components must involve international collaboration. Robot components must be interchangeable and usable off the shelf. Standards for robot–robot and human–robot interactions will have to be developed.

5.5 Robotic Advances

5.5.1 Sensing

- **Short-term (5 years)**: Gradual replacement of special hardware; 3D vision sensors in low resolution; sensor fusion is task-specific and relies on calibration; limited by processing power; use of attention mechanisms.
- **Mid-term (10 years)**: High frame rate of visual sensors; greatly improved 3D vision sensors; no moving parts in laser scanners; advanced task-dependent sensor fusion; multiple sensor modalities; step change in visual serving; known events interpreted.
- **Long-term (15 years)**: Visual processes on sensor or dedicated processors; multi-modal sensing for intrinsic safety; sensing on chip; perception techniques take over from fusion (closer to human perception system); no longer task-dependent.

5.5.2 *Hardware and Software Design*

- **Short-term (5 years)**: Hierarchical architectures running on a single system; architecture may use multiple cores for specific purpose; separate tools exist to aid the design of the aspects of robot and application; simplistic models, which cannot be linked; shape memory alloys and electro-active polymers for microrobots; some use of carbon/composite/metal foams; lack of standards for model descriptions; simulation not as good as real-world experiments; long computation times.
- **Mid-term (10 years)**: Hybrid or layered service-oriented architectures; loosely coupled distributed modules (real-time agents); integrated tool chain for design of robot and application (easily extendable); dynamic robot models; shape memory alloys and electro-active polymers for robot reconfiguration; biomimetic/sensing materials; some use of nano-materials; standard language for model description; interchangeable models; modeling of flexible and soft bodies; improved cybernetics.
- **Long-term (15 years)**: Component compositionality and self-configuration; globally distributed, resource-aware architectures; integrated tools chain to custom-build robots; detailed, easy-to-use dynamic models for robot and environment; increased use of nano-materials; use of biomimetic materials and biological tissue; intelligent materials and structures; real-time, dynamic modeling and interpretation allow for the accurate assessment of the robot's and the world's state.

5.5.3 *Planning and Control*

- **Short-term (5 years)**: Manual programming superior to automated planning (optimized process path based on human experience); randomized motions as planning alternative; control through cascades; state–space controller; sliding mode controller; feedback linearization.
- **Mid-term (10 years)**: Automated mission and process planning using databases of expert knowledge; predictive, distributed, self-calibrating, self-tuning controllers.
- **Long-term (15 years)**: Autonomous, online planning for tasks of high dimensionality; learn from humans (often interactively); fault-tolerant controllers; automatic reconfiguration of controllers.

5.5.4 *Cooperating Robots and Intelligence*

- **Short-term (5 years)**: Teams of robots; centralized control and communications; tasks specified for each individual robot; use of a common map; parts of robot systems use learning methods; well-defined conditions; learning from expert teacher; task-specific end effectors; mostly pre-programmed or taught grasping strategies; flexibility with tool changers.

- **Mid-term (10 years)**: Distributed control; inter-agent communication; task specified for team; games and swarm theories are applied; essential parts of controllers use learning methods; learning by experience; learning by demonstration; multi-finger grippers for a variety of objects; grasps computed online; gripping of human tools.
- **Long-term (15 years)**: Cooperation without explicit representation of action; skill-based or learning-based automation; complete robotics systems use learning methods (learning by observation, flexible conditions); dexterous hands; grasping of all objects; use of multiple hands; human dexterity and assembly skills.

5.5.5 Real-Time Communication and Human–Machine Interface

- **Short-term (5 years)**: Numerous specialized protocols; Ethernet-based communication starts to take over as de facto standard; mostly graphical or text-based interfaces; few haptic devices and use of human interaction channels; touch interfaces.
- **Mid-term (10 years)**: New protocols using ontologies, logic, probabilistic or geometric models, rule sets; human interaction channels, which humans have to learn; some telepresence; haptic input devices; learning interfaces.
- **Long-term (15 years)**: Components can figure out each other's protocols; components negotiate required quality of service; interaction using human channels utilizing cognitive approaches; neural interfaces; non-invasive brain interfaces.

5.5.6 Energy Management and Safety

- **Short-term (5 years)**: Mostly electric, pneumatic or hydraulic motors; lightweight high-density actuators; standard gears; mostly external power or local storage; regenerative brakes available, but not used often; sensor-based physical safety; hardware safety through redundancy; software safety through formal approaches to programming.
- **Mid-term (10 years)**: Continuously variable transmissions; ball–socket joints; improved energy saving and power–weight ratio; local energy conversion/generation; regeneration is standard; planners conserve energy; model-based hardware and software failure detection and isolation; application safety (explosives, food, medicine, etc.).
- **Long-term (15 years)**: High-energy efficiency; safe, powerful actuators; micro actuation; use of smart materials; powerful pneumatics and hydraulics; efficient

wireless power transfer; system efficiency continues to increase; predictive failure detection; safe automatic obstacle avoidance; detection of the intention of a person.

5.5.7 Navigation and Locomotion

- **Short-term (5 years)**: Engineering solutions to locomotion; locomotion inside the human body through external force fields; navigation expensive (computation and sensors); localization and mapping in controlled environments solved.
- **Mid-term (10 years)**: Biomimetic locomotion in/on water and on land; bipedal locomotion in structured environments; some perception-based localization; simultaneous localization and mapping for challenging environments; collision avoidance considers dynamic objects.
- **Long-term (15 years)**: Bipedal locomotion in unstructured environments (mostly indoors); energy efficiency; autonomous in-body locomotion; simultaneous localization and mapping in unconstrained environments; collision avoidance with dynamic, non-cooperative obstacles through perception.

5.6 Service Robots Forecast

5.6.1 Professional Service Robots

The evolution in sales of professional service robots from 2003 to 2008 (Karlsson 2004; IFR – Statistical Department 2009) is shown in Table 5.1. The sales data are

Table 5.1 Sales numbers of professional service robots in 2003, 2007 and 2008

Type of robot	Sales in 2003	Sales in 2007	Sales in 2008
Field robotics	110	3,603	4,959
Professional cleaning	212	156	156
Inspection and maintenance systems	18	190	170
Construction and demolition	225	285	362
Logistics systems	28	475	533
Medical robotics	218	651	867
Defense, rescue and security applications	319	4,581	6,274
Underwater systems	496	155	182
Mobile platforms in general use	262	272	352
Public relations robots	6	25	41
Total number of units	**2,302**	**10,395**	**13,904**

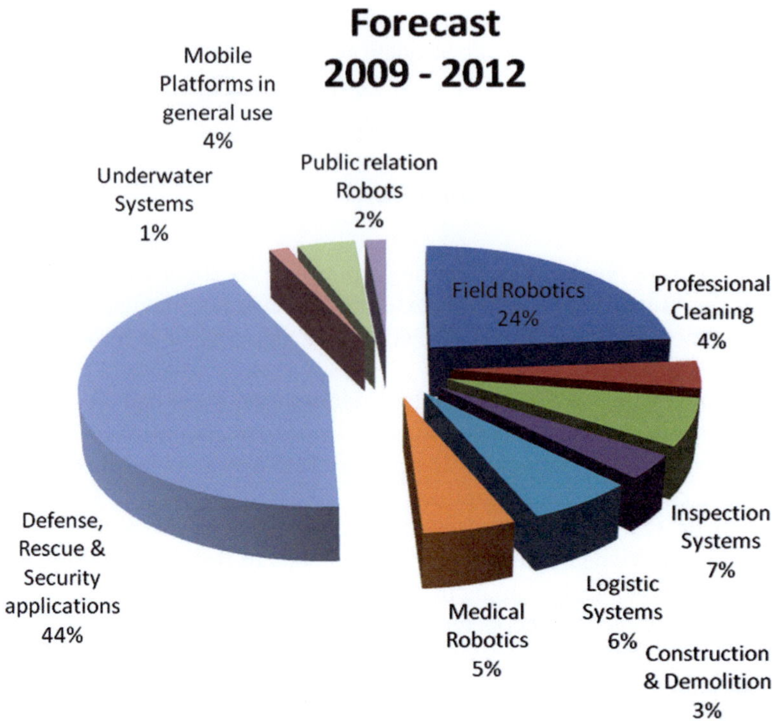

Fig. 5.4 Professional service robots sales forecast (2009–2012)

divided into types of professional service robots, stand out sales for the field and the defense followed by the logistics and healthcare.

Owing to the economic crisis that began in 2007, the sales forecast of professional service robots will suffer until 2012 (when it is estimated that all markets will stabilize). This sales forecast can be observed in Fig. 5.4 as a percentage for each type of robots. The number of robots to be sold in the period 2009–2012 is estimated to be 49,415 units, which will involve an expenditure of $9.87 billion.

5.6.2 Domestic Service Robots

The evolution in sales of domestic service robots from 2003 to 2008 (Karlsson 2004; IFR – Statistical Department 2009) is shown in Table 5.2. The sales data are divided into types of domestic service robots, especially the sales for domestic tasks (more specifically in the vacuum and floor cleaning sector).

The sales forecast for 2009–2012 is shown in Fig. 5.5. A high number of sales of household robots is expected thanks to cheaper robotic devices and public acceptance. Educational robots and interactive toys will also increase markedly.

Table 5.2 Sales for domestic service robots in 2003, 2007 and 2008

Type of robot	Sales in 2003	Sales in 2007	Sales in 2008
Robots for domestic tasks	397,500	660,536	966,968
Entertainment robots	136,968	929,902	785,922
Handicap assistance	65	2	
Personal transportation	184		
Home security and surveillance		258	300
Total number of units	**534,717**	**1,590,698**	**1,753,190**

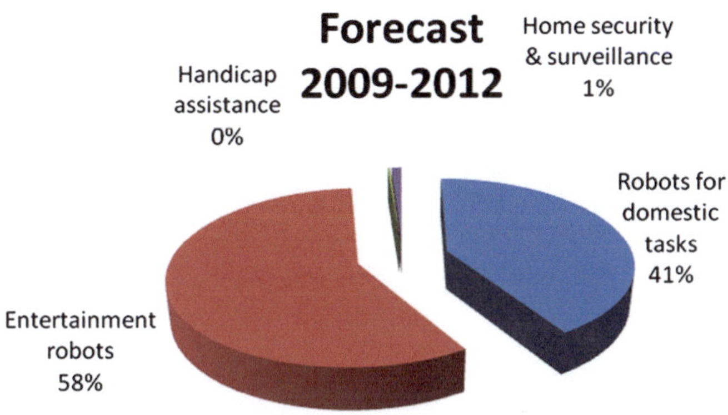

Fig. 5.5 Domestic service robots sales forecast (2009–2012)

There will be a tendency towards robotic maids to care for elderly or disabled persons and for surveillance and home security. The number of robots to be sold in the period 2009–2012 is estimated to be 11,591,700 units, which will involve an expenditure of $3.1 billion.

5.7 Home Automation and the Digital Home

5.7.1 History

As part of the development of western society, the family model is made up of a marriage with children, house, garage, kitchen, lounge, toilets, bathrooms and bedrooms. Thus, it is logical that since the 1950s people have fantasized about the idea of implementing all the developments achieved in the production industry and transportation into the automation of homes.

The origin of home automation can be traced to the 1970s with the appearance of the X10 protocol. In the 1980s and 1990s, many non-residential constructions were built (they were called intelligent buildings). The intelligent building model failed to materialize in residential construction. Construction techniques in electrical, plumbing, sanitary and air conditioning installations have not changed fundamentally since the 1950s except for the provision of radio, telephone, TV, personal computers, mobile phones and Internet access at home. This failure is because economies of scale have not been reached and technology standards have not been agreed, which has fostered a high price.

The digital home is seated on four pillars (CEDOM – Home Automation Spanish Association 2010):

- Comfort: local and remote control lights and blinds on timers.
- Safety: safety of goods and people. Technical alarms for flooding, fire, smoke, gas leak and intrusion.
- Leisure and communications: telephony, Internet, multimedia, data, TV.
- Energy efficiency: controlling power consumption, gas, water and climate control.

Although in non-residential construction it is common to find a control data network, a multimedia network in residential construction is still hard to find. The application of home automation is considered exclusive and expensive, but in the past 2 years great strides have been made with the popularization of KONNEX.

5.7.2 *Market Trends*

The current status of the construction market has led to a fall in related sales. Thus, home automation has been affected, slowing the trend of exponential growth seen in recent years. Table 5.3 shows a sales forecast for digital homes for 2011 in Spain.

The development of home automation will come with Internet access, which will encourage the use of new technologies such as:

- Semantic Web
- Internet of Things
- Augmented Reality
- Smart Grids

Figure 5.6 shows the percentage of households (Eurostat 2010) with Internet access in different states of the European Union and the rest of the world.

Table 5.3 Projected sales for 2011 based on the value of home automation installations

Type of home automation installation	Number of houses
Basic (1.000€)	2,000
Medium (3.000€)	1,000
Premium (12.000€)	200
Luxury (20.000€)	150

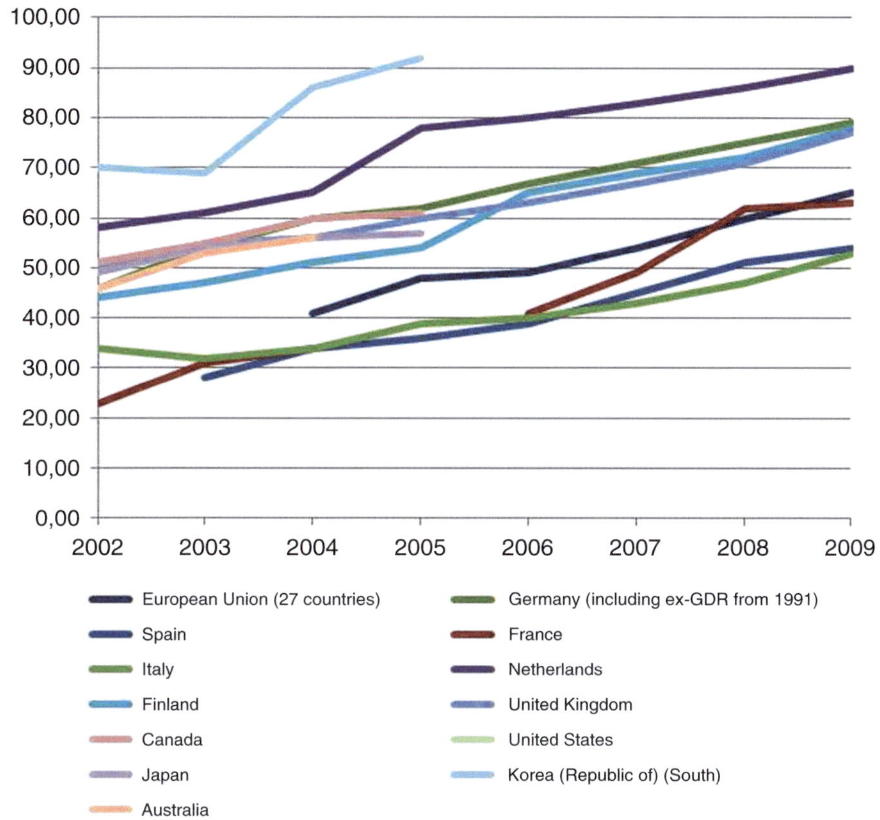

Fig. 5.6 Percentage of homes with access to the Internet

5.8 Home Automation and Service Robots

5.8.1 Baseline

Home Automation and robotics are immature markets regulated more by supply than by demand requirements. Due to the idiosyncrasy of these markets, robotic and home automation markets share a common feature: uncertainty. This factor makes palpable changes in development times of high productivity and social recognition, in contrast with other days of stagnation.

At present, in a global crisis, industry estimates are up in the air. Estimates have had to be revised downwards. However, this same situation offers two great opportunities to develop markets and capacity for growth. The market for service robots has undoubtedly a long-distance business in which functionality, technical training, R&D and marketing should work well to bring the desired and final release.

So far, all comments are valid for both markets – both in full swing and independent of each other. However, right now there are professionals in both sectors that offer an intimate union between automation and service robotics in search of fringe benefits and commonalities.

In the field of automation, technology is at a point of ripeness. Proprietary protocols and even de facto standard features offer potential growth rather than proven. Yet the public does not perceive the value it can bring by introducing new technologies into their buildings and homes. Services can be aimed at improving the quality of life, safety of goods and people, energy efficiency and recreational features to enjoy in our homes, public places or buildings.

Service robotics is at an earlier stage in terms of market maturity and home automation. Although the channels and equipment are gradually reaching stores, the majority of the proposals are seen as experimental products that are only presented at trade fairs and congresses. This is a major obstacle to the desired commercial development.

5.8.2 Trends

As discussed in the previous section, the merging of robotics and automation would offer more potential for further development than separately. Belonging to the technology industry, both are in earlier maturing markets and both have great technology. Factors to be taken into account in the market for robotics and automation are:

- The global financial crisis that began in 2009
- The massive use of technology in everyday life
- The lack of power supply for current demand
- The gradual, inexorable introduction of technological equipment in buildings and homes
- Public and private improvements in the processes of research and development
- A large, value-added financial contribution to development tasks, introducing changes to the laboral model.
- Achieving interoperability between the above mentioned markets offers several advantages:
- Service robotics add a closer and more usable interface to home automation, thus removing one of the most important handicaps of current home automation, where the user has a frosty relationship with their Smart Home system, often limited to touch screens and switches.
- The introduction of robotics into the home, with the help of the home automation system, fosters a social concept of closeness and accelerates its growth in the market.
- Service robotics will leverage its impact in the market as a means for turning itself into a popular technology, available to everybody and properly using the services offered by home automation systems.

5.8.3 Home Automation, Robotics and the Healthcare Sector

The model will develop hand in hand with the introduction of automation in the management of the building, efficiently managing energy consumption, monitoring the vital signs of patients, offering priority access and activities, maximizing resources and improving the connectivity among patients, doctors and families.

Robotics can pass information to patients, relatives and medical guides. Activities such as accompanying the sick and elderly, creating dynamic and enabling activities and stimulants, providing psychomotor and motor exercises for patients with these needs, distributing food, generating menus, cooking, cleaning are all closely linked to the direct interaction with intelligent man.

5.8.4 Home Automation, Robotics and the Leisure Sector

Home automation is fast becoming one of the factors in the development in this sector. We are consuming more leisure in our homes and public buildings, which will inevitably introduce a barrage of technology to facilitate and implement those needs. These include video game consoles, the introduction of non-traditional sectors of these markets (older adults), increasing and improving the quality of Internet connections, the sensations offered by the new interfaces of the equipment, mass consumption and the increase in these electronics.

Service robotics has much yet to say in the leisure sector. Despite being penalized by the overhead costs of the equipment, a leading position in which the robots will become our best friends to enjoy leisure time with is slowly taking shape. This will lead to improved user interaction and an empathy and enjoyment of the game.

5.8.5 Home Automation, Robotics and Senior Citizens

Home automation has become an interesting model in which public and private entities telecare our elders. In an aging society, automation is providing an extra service that you can now enjoy. The guardianship of elderly people in their homes to improve their independence – through videoconferencing telemedicine, patterning and routines to provide proactive assistance services and interacting with friends and family – are other areas where there is more work needed.

Service robotics offers a world of possibilities. Along with automation, the introduction of services is multiplied exponentially, with assisting disabled persons, carrying out routine household tasks, generating alarms and social support some of the most important.

5.8.6 Home Automation, Robotics and Disabled Persons

Home automation is experiencing a tremendous growth in improving the quality of life and autonomy of people with disabilities. However, the future is also bright in areas such as enhancing autonomy in public buildings, customizing home adaptations, safety management, increased user mobility, connectivity, medical or the generation of custom adaptations each user.

The robotics field offers a great development including interaction with people, social support, performing routine tasks at home, mobility of people, generating alarms or increased autonomy of individuals.

5.8.7 Home Automation, Robotics and Public Environments

Here, we are talking about public spaces such as streets, parks and gardens where robots provide integrated security services, maintenance, the intelligent and integrated management of traffic, announcements and notices and the integration of services. In addition, this includes public buildings such as museums or city halls, which implement energy efficiency criteria, access management and guidance and support visits.

Glossary

4D/RCS Reference model architecture for military unmanned vehicles on how their software components should be identified and organized.

Anykode Marilou Modeling and simulation environment for mobile robots, humanoids, articulated arms and parallel robots operating in real-world conditions that respect the laws of physics.

Architectural patterns Concept that solves and delineates some essential cohesive elements of a software architecture.

Autonomous The capacity of a rational individual to make an informed, uncoerced decision.

Bundles Normal jar components with extra manifest headers in OSGi protocol.

Carmen Carnegie Mellon Robot Navigation Toolkit acronym – this is an open source toolkit for mobile robot control.

CLARAty Integrated framework for reusable robotic software.

CORBA Common Object Request Broker Architecture is the standard defined by the Object Management Group that enables software components written in multiple computer languages and running on multiple computers to work together.

Development platform Software design and programming of a system or software package.

DH Compliant Interoperability protocol between robotics and home automation systems developed by the DHCompliant consortium.

Digital home An extension of home automation concept adding TI.

DLNA Digital Living Network Alliance – a non-profit collaborative trade organization and a protocol for audio/video media exchange.

GENA General Event Notification Architecture, HTTP notification architecture that transmits notifications between HTTP resources.

HTTP Hypertext Transfer Protocol, networking protocol for distributed, collaborative, hypermedia information systems.

HTTPMU Variant of HTTPU that uses IP multicast.

HTTPU Extension of the HTTP/1.1 protocol using UDP as the data transport instead of the usual TCP protocol.

I.G. Alonso et al., *Service Robotics within the Digital Home*, Intelligent Systems, Control and Automation: Science and Engineering 53, DOI 10.1007/978-94-007-1491-5,

© Springer Science+Business Media B.V. 2011

Industrial Robot Automatically controlled, reprogrammable, multipurpose manipulator programmable in three or more axes.

Industrial Safety Set of safety conditions for workers, materials and tools in an industrial environment.

Interoperability The ability of diverse systems and organizations to work together (inter-operate).

JAUS Joint Architecture for Unmanned Systems, which was originally an initiative by the US Department of Defense.

JINI Network architecture for the construction of distributed systems in the form of modular cooperating services.

Logistics The management of the flow of the goods, information and other resources in a repair cycle between the points of consumption to meet the requirements of costumers.

MARIE Mobile and Autonomous Robotics Integration Environment acronym, development environment for robotics platforms integration.

Microsoft Robotics Studio Development of robotic platform point to point. Visual programming tool focused to develop and debug robotics applications.

Middleware Computer software that connects software components or people and their applications

Miro Middleware for Mobile Robot Applications.

NATO STANAG 4586 NATO Compliant Ground Control System for UAV.

NS-2/NS-3 Network Simulator, discrete event network simulator.

OMNET Component-based, modular and open architecture discrete event network simulator.

OpenRDK Modular software framework focused on rapid development of distributed robotic systems.

OPNET Optimized Network Engineering Tool, software tool for network modeling network performance.

Orca Open source framework for developing component-based robotic systems.

OROCOS Open Robot COntrol Software acronym –a project that provides a free software toolkit for real-time robot arm and machine tool control.

OSGi Open Services Gateway Initiative acronym – a module system and service platform for the Java programming language that implements a complete and dynamic component model.

Personal Robotics Assistants Robots in charge of a person's healthcare and assistance in servant or professional labors.

Player/Stage/Gazebo Free software robot interface and simulation system.

Professional Service Robots Branch of service robots entrusted to provide assistance with work.

RMI Remote Method Invocation provides a simple and direct model for distributed computation with Java objects so that one can write distributed.

Robotic Kits Building a robot from a kit is one of the best ways to get involved with robotics.

Robot toys Robots for entertainment.

Salutation Technique for identifying resources.

Semantic Web services (SWS) Web services with semantic information.

Service robotics Science that studies service robots.

Service robots Robots that operate semi or completely autonomously to realize useful services for humans, excluding manufacturing.

Simulation platform Software that allows the study of systems behavior before implementation in a real environment.

Smart home Synonym for the digital home.

SOAP Simple Object Access Protocol is a protocol specification for exchanging structured information in the implementation of WS in computer networks.

SSDP Simple Service Discovery Protocol is a network protocol based on the Internet Protocol Suite for the advertisement and discovery of network services and presence information.

SysML Systems Modeling Language is a general purpose modeling language for systems engineering applications.

TCP Transmission Control Protocol.

Teleassistance Healthcare, in their own house, of elderly persons or with some dependence degree.

Telepresence A set of technologies that allow a person to feel as if they were present, to give the appearance that they were present or to have an effect, via telerobotics, at a place other than their true location.

UDP User Datagram Protocol.

UPnP Universal Plug and Play, a set of networking protocols for primarily residential networks without enterprise class devices that permits networked devices.

Urbi Universal Robot Body Interface, open source cross-platform software platform in C++ used to develop applications for robotics and complex systems.

USARSim Open source high fidelity robot simulator that can be used for both research and education.

Web services Software system designed to support interoperable machine-to-machine interaction over a network.

Webots Professional robot simulator.

XML Extensible Markup Language is a set of rules for encoding documents in machine-readable form.

References

Journal Articles

Albus J, Huang H, Messina E (2002a) 4D/RCS A reference model architecture for unmanned vehicle systems, version 2.0, NISTIR 6910, National Institute of Standards and Technology, Gaithersburg, MD

Gomez H (2008) La base instalada de PC sobrepasa los 1.000 millones de unidades. Cio España. 24/06/2008. http://www.idg.es/cio/La_base_instalada_de_PC_sobrepasa_los_1.000_millones_de_unidades/doc69106-mercado.htm

Pransky J (1996) Service robots. How should we define them? Serv Robot Int J 2(1):4–5

Journal Articles by DOI

Ahn Sang Chul, Kim Jin Hak, Lim Kiwoong, Ko Heedong, Kwon Yong-Moo, Kim Hyoung-Gon (2005) UPnP approach for robot middleware. In: IEEE International Conference on Robotics and Automation (ICRA), 2005, Barcelona, SPAIN, pp 1959–1963. doi: 10.1109/ROBOT.2005.1570400

Albus J, Murphy K, Lacaze A, Legowik S, Balakirsky S, Hong T, Shneier M, Messina E (2002b) 4D/RCS sensory processing and world modeling on the Demo III experimental unmanned ground vehicles. In: Proceedings of the 2002 IEEE international symposium on intelligent control, pp 885–890. doi: 10.1109/ISIC.2002.1157879

Allard J, Chinta V, Gundala S, Richard GG (2003) Jini meets UPnP: an architecture for Jini/UPnP interoperability. In: Proceedings of the 2003 symposium on applications and the internet, 2003, Orlando, FL, pp 268–275. doi: 10.1109/SAINT.2003.1183059. http://www.computer.org/portal/web/csdl/doi?doc=abs/proceedings/saint/2003/1872/00/18720268abs.htm

Arnold K (1999) The Jini architecture: dynamic services in a flexible network. In: Proceedings of the 36th annual ACM/IEEE design automation conference, New Orleans, LA, USA, pp 157–162. doi: 10.1145/309847.309906

Bajaj S, Breslau L, Estrin D, Fall K, Heidemann J, Huang P, Kumar S, McCanne S, Rejaie R, Sharma P, Varadhan K, Xu Y, Yu H, Zappala D (1999) Improving simulation for network research. IEEE Computer (To appear, a preliminary draft is currently available as USC technical report pp 99–702). doi: 10.1.1.44.7229

Ballantyne GH (2002) Robotic surgery, telerobotic surgery, telepresence, and telementoring. Surg Endosc 16(10):1389–1402. doi: 10.1007/s00464-001-8283-7

Begum A, Lee M, Kim YJ (2010) A simple visual servoing and navigation algorithm for an omnidirectional robot. In: 3rd international conference on human-centric computing (HumanCom), 2010, Cebu, Philippines, pp 1–5. doi: 10.1109/HUMANCOM.2010.5563325

Cheng-Wei Chen, Chung-Kai Chen, Jyh-Cheng Chen, Chien-Tan Ko, Jenq-Kuen Lee, Hong-Wei Lin, Wang-Jer Wu (2004) Java RMI over heterogeneous wireless networks. IEEE Int Conference Commun (ICC) 3:1391–1395. doi: 10.1109/ICC.2004.1312740 DOI: dx.doi.org

Coulson G, Blair G, Clarke M, Parlavantzas N (2002) The design of a configurable and reconfigurable middleware platform. Distrib Comput 15(2):109–126. doi: 10.1007/s004460100064

Cousins S (2010) ROS on the PR2 [ROS topics]. IEEE Rob Autom Mag 17(3):23–25. doi: 10.1109/MRA.2010.938502

Curbera F, Duftler M, Khalaf R, Nagy W, Mukhi N, Weerawarana S (2002) Unraveling the Web services web: an introduction to SOAP, WSDL, and UDDI. IEEE Internet Comput 6(2):86–93. doi: 10.1109/4236.991449

Ferguson P, Huston G (1998) What is a VPN? http://citeseerx.ist.psu.edu/viewdoc/summary? doi: 10.1.1.28.972

Furmento N, Lee W, Mayer A, Newhouse S, Darlington J (2002) ICENI: an open grid service architecture implemented with Jini. In: SC conference, 0:37. IEEE Computer Society, Los Alamitos, CA. doi: http://doi.ieeecomputersociety.org/10.1109/SC.2002.10027

Furmento N, Hau J, Lee W, Newhouse S, Darlington J (2004) Implementations of a service-oriented architecture on top of Jini, JXTA and OGSI. In: Grid computing, 3165:249–261. Lecture notes in computer science. Springer, Berlin/Heidelberg. http://dx.doi.org/10.1007/978-3-540-28642-4_11

Graf B, Hans M, Schraft RD (2004) Care-O-bot II—development of a next generation robotic home assistant. Auton Robots 16(2):193–205. doi: 10.1023/B:AURO.0000016865.35796.e9 DOI: dx.doi.org

Gu T, Pung HK, Zhang DQ (2004) Toward an OSGi-based infrastructure for context-aware applications. IEEE Pervasive Comput 3(4):66–74. doi: 10.1109/MPRV.2004.19

Gupta R, Talwar S, Agrawal DP (2002) Jini home networking: a step toward pervasive computing. Computer 35(8):34–40. doi: 10.1109/MC.2002.1023786

Hayes AT, Martinoli A, Goodman RM (2003) Swarm robotic odor localization. In: Proceedings of the IEEE/RSJ international conference on intelligent robots and systems (IROS'01), Sendai. doi: 10.1109/IROS.2001.976311

Jackson J (2007) Microsoft robotics studio: a technical introduction. IEEE Rob Autom Mag 14:82–87. doi: 10.1109/M-RA.2007.905745

Jones JL (2006) Robots at the tipping point: the road to iRobot Roomba. IEEE Rob Autom Mag 13(1):76–78. doi: 10.1109/MRA.2006.1598056

Kang DO, Kang K, Choi S, Lee J (2005) UPnP AV architectural multimedia system with a home gateway powered by the OSGi platform. IEEE Trans Consum Electron 51(1):87–93. doi: 10.1109/TCE.2005.1405704

Kawamura K, Pack RT, Bishay M, Iskarous M (1996) Design philosophy for service robots. Rob Autom Syst 18(1–2):109–116. doi: 10.1016/0921-8890(96)00005-X DOI: dx.doi.org

Kerstin S-E, Green A, Hüttenrauch H (2003) Social and collaborative aspects of interaction with a service robot. Rob Autom Syst 42(3–4):223–234. doi: 10.1016/S0921-8890(02)00377-9 DOI: dx.doi.org

Kim D, Lee J, Kwon WH, Yuh IK (2002) Design and implementation of home network systems using UPnP middleware for networked appliances. IEEE Trans Consum Electron 48:963. doi: 10.1109/TCE.2003.1196427

Knoop S, Vacek S, Zöllner R, Au C, Dillmann R (2004) A CORBA-based distributed software architecture for control of service robots. Proc Int Conf Intell Robots Syst 4:3656–3661. doi: 10.1109/IROS.2004.1389983 DOI: dx.doi.org

Koenig N, Howard A (2004) Design and use paradigms for gazebo, an open-source multi-robot simulator. In: IEEE/RSJ international conference on intelligent robots and systems, Taipei. doi: 10.1109/IROS.2004.1389727

Koren Y, Borenstein J (1991) Potential field methods and their inherent limitations for mobile robot navigation. IEEE Int Conf Robot Automation 2:1398–1404. doi: 10.1109/ROBOT.1991.131810

Lara R, Roman D, Polleres A, Fensel D (2004) A conceptual comparison of WSMO and OWL-S, vol 3250. Springer, Berlin, pp 254–269. doi: 10.1007/978-3-540-30209-4_19

Mamen R (2003) Applying space technologies for human benefit; the Canadian experience and global trends. In: Proceedings of international conference on recent advances in space technologies, RAST '03, Istanbul, pp 1–8. doi: 10.1109/RAST.2003.1303381

Marco TG, Cristina V, Paolo A, Luca P, Dario GA, Massimo P (2010) Design considerations about a photovoltaic power system to supply a mobile robot. In: IEEE international symposium on industrial electronics (ISIE), Bari, 2010, pp 1829–1834. doi: 10.1109/ISIE.2010.5637724

Matsumaru T (2009) Discrimination of emotion from movement and addition of emotion in movement to improve human-coexistence robot's personal affinity. In: The 18th IEEE international symposium on robot and human interactive communication, RO-MAN 2009, Toyama, pp 387–394. doi: 10.1109/ROMAN.2009.5326345

Michel O (1998) Webots: symbiosis between virtual and real mobile robots. Lecture Notes in Computer Science. doi: 10.1007/3-540-68686-X_24

Michel O (2004) Cyberbotics Ltd. Webots TM: professional mobile robot simulation. Int J Adv Robotic Syst 1(1):39–42. doi: 10.1.1.86.1278

Miller B, Pascoe R (1999) Mapping salutation architecture APIs to bluetooth service discovery layer. Bluetooth Consortium 1.C.118/1.0 1. http://citeseerx.ist.psu.edu/viewdoc/summary?doi: 10.1.1.39.1063

Miller BA, Nixon T, Tai C, Wood MD (2001) Home networking with universal plug and play. IEEE Commun Mag 39(12):104–109. doi: 10.1109/35.968819

Mohammed N, Al-Jaroodi J (2008) Characteristics of middleware for networked collaborative robots. In: International symposium on collaborative technologies and systems, CTS 2008, Irvine. doi: 10.1109/CTS.2008.4543973

Mok SM, Wu C (2006) Automation integration with UPnP modules, 5 pp. doi: 10.1109/DELTA.2006.18

Morgan S (2000) Jini to the rescue [computer network interconnection technology]. IEEE Spectr 37(4):44–49. doi: 10.1109/6.833027

Nesnas I, Wright A, Bajracharya M, Simmons R, Estlin T, Kim WS (2003) CLARAty: an architecture for reusable robotic software. In: SPIE Aerosense Conference, Orlando. doi: 10.1117/12.497223

Oh Yeon-Joo, Lee Hoon-Ki, Paik Eui-Hyun, Park Kwang-Roh, Kim Jung-Tae (2007) Implementation of the DLNA Proxy System for sharing home media contents. IEEE Trans Consum Electron 53(1):139–144. doi: 10.1109/TCE.2007.339515

Sakagami Y, Watanabe R, Aoyama C, Matsunaga S, Higaki N, Fujimura K (2002) The intelligent ASIMO: system overview and integration. In: IEEE/RSJ international conference on intelligent robots and systems, 2002, Lausanne, vol 3, pp 2478–2483. doi: 10.1109/IRDS.2002.1041641

Shoemaker CM, Bornstein JA (1998) The Demo III UGV program: a testbed for autonomous navigation research. In: Intelligent control (ISIC).In: Proceedings held jointly with IEEE international symposium on computational intelligence in robotics and automation (CIRA), Houghton, intelligent systems and semiotics (ISAS), Gaithersburg, pp 644–651. doi: 10.1109/ISIC.1998.713784, doi: dx.doi.org

Suri N, Bradshaw JM, Carvalho MM, Cowin TB, Breedy MR, Groth PT, Saavedra R (2003) Agile computing: bridging the gap between grid computing and ad-hoc peer-to-peer resource sharing. In: Proceedings of 3rd IEEE/ACM international symposium on cluster computing and the grid, CCGrid 2003, Tokyo, pp 618–625. doi: 10.1109/CCGRID.2003.1199423. http://ieeexplore.ieee.org/Xplore/login.jsp?url=http%3A%2F%2Fieeexplore.ieee.org%2Fiel5%2F8544%2F27003%2F01199423.pdf%3Farnumber%3D1199423&authDecision=-203

Takeda K, Nasu Y, Capi G, Yamano M, Barolli L, Mitobe K (2001) A CORBA-based approach for humanoid robot control. Ind Robot Int J 28(3):242–250. doi: 10.1108/01439910110389407 DOI: dx.doi.org

Vinoski S (1997) CORBA: integrating diverse applications within distributed heterogeneous environments. IEEE Commun Mag 35:46–55. doi: 10.1109/35.565655 DOI: dx.doi.org

Volpe R, Nesnas I, Estlin T, Mutz D, Petras R, Das H (2001) The CLARAty architecture for robotic autonomy. In: Aerospace conference, 2001, IEEE Proceedings, Big Sky. doi: 10.1109/AERO.2001.931701
Wang LF, Tan KC, Prahlad V (2000) Developing Khepera robot applications in a Webots environment. In: Proceedings of 2000 international symposium on micromechatronics and human science, MHS 2000, Nagoya. doi: 10.1109/MHS.2000.903293
Westhoff D, Scherer T, Stanek, H, Zhang J, Knoll A (2004) A flexible framework for task-oriented programming of service robots. IN VDI BERICHTE, 1841. doi: 10.1109/ICIA.2005.1635052
Woo E, MacDonald BA, Trepanier F (2003) Distributed mobile robot application infrastructure. In: Proceedings of international conference on intelligent robots and systems, Las Vegas, vol 2, pp 1475–1480. doi: 10.1109/IROS.2003.1248852 DOI: dx.doi.org
Wooden D, Malchano M, Blankespoor K, Howardy A, Rizzi AA, Raibert M (2010) Autonomous navigation for BigDog. In: IEEE international conference on robotics and automation (ICRA), 2010, Anchorage, pp 4736–4741. doi: 10.1109/ROBOT.2010.5509226
Zeng X, Bagrodia R, Gerla M (1998) GloMoSim: a library for parallel simulation of large-scale wireless networks. In: Proceedings of twelfth workshop on parallel and distributed simulation, PADS '98, Banff. doi: 10.1109/PADS.1998.685281
Zhang Z, Cao Q, Zhang L, Lo C (2009) A CORBA-based cooperative mobile robot system. Ind Robot Int J 36(1):36–44. doi: 10.1108/01439910910924657
Zlajpah L (2008) Simulation in robotics. Math Comput Simul 79(4):879–897. doi: 10.1016/j.matcom.2008.02.017

Books

Bartlett N (2009) OSGI in practice. Amazon.com
Berners-Lee T, Hendler J, Lassila O (2001) The semantic web. A new form of Web content that is meaningful to computers will unleash a revolution of new possibilities. Scientific American Magazine, May 17, 2001
Bischoff R, Guhl T (2009) Robotic visions to 2020 and beyond – The strategic research agenda (SRA) for robotics in Europe. www. robotics-platform. eu/sra. Last accessed on 23 Mar 2010
Bottazzi S, Caselli S, Reggiani M, Amoretti M (2002) A software framework based on real-time CORBA for telerobotic systems. In: IEEE international conference on intelligent robots and systems, Lausanne, pp 3011–3017
Bräunl T, Graf B (2008) Embedded robotics – mobile robot design and applications with embedded systems, 3rd edn. Springer-Verlag Berlin Heidelberg
Breazeal Cynthia L (2004) Designing sociable robots. MIT Press, Cambridge, MA/London
Chen H (2000) Developing a dynamic distributed intelligent agent framework based on the Jini architecture, Master's thesis, University of Maryland, Baltimore County (January 2000)
Collaborative T (2009) From internet to robotics. http://citeseerx.ist.psu.edu/viewdoc/download?doi=10.1.1.159.4278&rep=rep1&type=pdf. Last accessed on 23 Mar 2010
Daconta MC, Obrst LJ, Smith KT (2003) The semantic web: a guide to the future of XML, Web services, and knowledge management. Wiley, Indianapolis
English W (2007) Joint Architecture for Unmanned Systems (JAUGS). Reference architecture specification, vol II, version 3.3
Everett HR, Laird RT, Carroll DM, Gilbreath GA, Heath-Pastore TA, Inderieden RS, Tran T, Grant KJ, Jaffee DM (2000) Multiple Resource Host Architecture (MRHA) for the Mobile Detection Assessment Response System (MDARS). SPAWAR Systems Technical Document 3026, Revision A
Gothing G, Hurdus J (2006) Implementation of JAUS on a 2004 CadillacSRX using a potential fields architecture. AUVSI's unmanned systems NorthAmerica. Orlando, FL

He H (2003) What is service-oriented architecture. webservices
Henning M, Vinoski S (1999) Advanced CORBA programming with C++. Addison-Wesley Professional, Reading, MA
IFR – Statistical Department (2009) World robotics 2009
IFR – Statistical Department (2010) World robotics 2010
Jeronimo M, Weast J (2003) UPnP design by example. A software developers guide to universal plug and play. Intel Press, Hillsboro
Karlsson J (2004) World robotics 2004. United Nations, Geneva
Keshav S (1988) REAL: A network simulator. Citeseer, University of California, Berkeley
OSGi Alliance (2003) OSGi service platform: the OSGi alliance. IOS Press, Amsterdam, The Netherlands
Schraft R-D, Schmierer G (2000) Service robots. A K Peters, Ltd, Natick
Spexard TP, Siepmann FHK, Sagerer G (2008) Memory-based Software Integration for Development in Autonomous Robotics. In: Burgard W, Dillmann R, Plagemann C, Vahrenkamp N (ed) Intelligent Autonomous Systems 10: IAS-10
Vajta L, Juhasz T (2005) The role of 3D simulation in the advanced robotic design, test and control, cutting edge robotics, Vedran Kordic, Aleksandar Lazinica and Munir Merdan (Ed.), ISBN: 3-86611-038-3, InTech

Online Documents (No DOI Available)

Allegro Software Development Corporation (2006) Networked digital media standards. A UPnP/DLNA overview, Massachusetts. http://www.allegrosoft.com/UPnP_DLNA_White_Paper.pdf. Accessed 12 Dec 2010
AnyKode (2010) anyKode Marilou – modeling and simulation environment for Robotics. http://www.anykode.com/index.php. Accessed 3 Dec 2010
Aquabot (2010) Aquabot official dealer site. http://www.aquabots.com/. Accessed 1 Dec 2010
ARA (Applied Research Associates) (2010). http://www.ara.com/. Accessed 2 Dec 2010
Arruda K (2008) DTCP-IP for DLNA
Australian Government – The Treasury (2002) A survey of international fiscal policy issues – current drivers and future challenges. http://www.treasury.gov.au/documents/382/HTML/docshell.asp?URL=04Fiscalarticle.html. Accessed 1 Dec 2010
Avancha S, Joshi A, Finin T (2001) Enhancing the bluetooth service discovery protocol, Techical Report TR-CS-01-08
Bagrodia R, Meyer R, Takai M, Chen Y, Zeng X, Martin J, Song HY (1998) Parsec: a parallel simulation environment for complex systems. IEEE Computer Society, Los Alamitos
Balaguer B, Balakirsky S, Carpin S, Lewis M, Scrapper C (2008) USARSim: a validated simulator for research in robotics and automation. In: Workshop on. “Robot Simulators”, Nice, France, Sept 2008
Beck D, Ferrein A, Lakemeyer G (2007) A simulation environment for middle-size robots with multi-level abstraction. In: Proceedings of the 2007 International RoboCup Symposium, Atlanta, USA
Blair GS, Coulson G, Robin P, Papathomas M (1998) An architecture for next generation mid-dleware. http://portal.acm.org/citation.cfm?id=1659232.1659249. Accessed 12 Dec 2010
Blind Driver Challenge (2010). National Federation of the Blind. http://www.blinddriverchallenge.org/bdcg/Default.asp. Accessed 3 Dec 2010
Boston Dynamics (2010) Boston dynamics: dedicated to the science and art of how things move. http://www.bostondynamics.com/robot_bigdog.html. Accessed 30 Nov 2010
Bray T, Paoli J, Sperberg-McQueen CM, Yergeau F, Maler E (2000) Extensible markup language (XML). http://www.w3.org/TR/REC-xml
Bray T, Paoli J, Sperberg-McQueen CM, Maler E, Yergeau F (2008) Extensible markup language (XML) 1.0, 5th edn. W3C Recommendation 26. http://www.w3.org/TR/2008/REC-xml-20081126/. Accessed 2 Dec 2010

Brown KL, Christianson L (2005) OPNET Lab manual
Bruyninckx H (2003) Open robot control software: the OROCOS project. In: Proceedings of the 2001 IEEE international conference on robotics and automation, ICRA 2001, Seoul, May 2001
Cañas JM, Matellán V, Montúfar R, Tonanzintla M (2006) Programación de robots móviles. Revista Iberoamericana de Automática e Informática Industrial, 3(2):99–110
Carnegie Mellon University Home Page (2010) http://www.cmu.edu/index.shtml. Accessed 20 Nov 2010
Carroll DM, Mikell K, Denewiler T (2004) Unmanned ground vehicles for integrated force protection. In: SPIE Proceedings of the 5422Cartif. http://www.cartif.com/. Accessed 2 Dec 2010
Cartif (2010) Fundación CARTIF – Centro Tecnológico Cartif. http://www.cartif.com/index.php/es/quienes-somos/centro-tecnologico-cartif.html. Accessed 1 Dec 2010
CDL Systems, NATO STANAG 4586 (2010) http://www.cdlsystems.com/index.php/stanag4586. Accessed 3 Dec 2010
CEDOM – Home Automation Spanish Association (2010) What is home automation? Systems that integrate with home automation. Home automation system (Spanish Website). http://www.cedom.es/que-es-domotica.php. Accessed 1 Dec 2010
Chang X (1999) Network simulations with OPNET. In: Proceedings of the 31st conference on winter simulation: simulationa bridge to the future, Phoenix, vol 1, pp 307–314
Chrysanthakopoulos G, Nielsen H (2007) Microsoft robotics studio: architecture overview [Online]. Available: http://www.Microsoft.Com/winme/0703/29490/Architecture_Overview/Local/index.html
Clark MN (2005a) JAUS implementation: robots gather for successful interoperability experiment. IEEE Rob Autom Mag. www.openjaus.com/. Accessed 2 Dec 2010
Clark MN (2005b) JAUS compliant systems offers interoperability across multiple and diverse robot platforms, unmanned systems North America conference, Baltimore, MD. www.openjaus.com/. Accessed 2 Dec 2010
Collett THJ, MacDonald BA, Gerkey BP (2005) Player 2.0: toward a practical robot programming framework. In: Proceedings of the Australasian conference on robotics and automation (ACRA), Sydney
CONSER (2002) Collaborative simulation for education and research (CONSER). http://www.isi.edu/conser/. Accessed 22 Nov 2010
Côté C, Létourneau D, Michaud F, Valin JM, Brosseau Y, Raievsky C, Lemay M, Tran V (2004) Code reusability tools for programming mobile robots. In: Proceedings of international conference on intelligent robots and systems, Sendai, Japan, vol 2, pp 1820 – 1825. doi: 10.1109/IROS.2004.1389661
Cyberbotics (2009) Webots reference manual release 6.1.5. http://www.cyberbotics.com/reference.pdf. Accessed 3 Dec 2010
DARPA (2007) DARPA urban challenge. http://www.darpa.mil/grandchallenge/index.asp. Accessed 3 Dec 2010
De la Pinta JR, Maestre JM, Camacho EF, Alonso IG (2011) Robots in the smart home: a project towards interoperability. International Journal of Ad Hoc and Ubiquitous Computing (IJAHUC). 7(3):192–201
Defense Update (2007) STANAG 4586 – NATO complient ground control system for UAV. International, Online Defense Magazine. http://defense-update.com/products/s/stanag_4586.htm. Accessed 3 Dec 2010
DH Compliant (2010) http://www.dhcompliant.com/. Accessed 2 Dec 2010
DLNA (2007) DLNA overview and vision whitepaper. http://www.dlna.org/news/DLNA_white_paper.pdf. Accessed 12 Dec 2010
Dobrescu R, Dobrescu M, Nicolae M (2007) Using UPnP services with an intelligent sensor network node. In: Proceedings of the 7th wseas international conference on applied informatics and communications, 2007. Vouliagmeni, GREECE, pp 373–376
Domotica Davinci (2010) http://www.domoticadavinci.com/. Accessed 2 Dec 2010
Erdei M, Wagner A, Sója K, Székely M (2001) A networked remote simulation architecture and its remote OMNeT ++ implementation. 15th European Simulation Multiconference (ESM 2001), CTU, Prague, Czech Republic. Modelling and simulation pp 235–242

Eurostat (2010) Households having access to the Internet at home. http://appsso.eurostat.ec.europa.eu/nui/show.do?dataset=isoc_pibi_hiac&lang=en. Accessed 1 Dec 2010

Fout T (2001) Universal plug and play in windows XP. https://www.isysguy.com/%5Cdocuments/Library/DataCom%5CUniversalPlugandPlayinWindowsXP.pdf. Accessed 12 Dec 2010

Fuentes S (2007) Qué es DLNA: a fondo. http://alatest.se/apps/reviews/21399060/-1/?ref=http%3A%2F%2Falatest.se%2Fprodukttester%2Fovriga-digitalkameror%2Fque-es-dlna-a-fondo%2Fpo3-83010536%2C257%2F. Accessed 12 Dec 2010

Fumio O, Junji O, Hideaki H, Hirokazu S (2004) Open robot controller architecture (ORCA). Nippon Kikai Gakkai Robotikusu, Mekatoronikusu Koenkai Koen Ronbunshu

Futurerobot (2010) Future robot. http://www.futurerobot.com/. Accessed 7 Dec 2010

General Atomics (2010) General atomics affiliates. http://www.ga.com/index.php. Accessed 1 Dec 2010

Gerkey BP, Vaughan RT, Howard A (2003) The player/stage project: tools for multi-robot and distributed sensor systems. In: Proceedings of the 11th international conference on advanced robotics ICAR 2003, Coimbra

Gostai (2010) The Urbi Software Development Kit. Version 2.0-49-g4358c61. Accessed 3 Dec 2010

Henderson TR, Roy S, Floyd S, Riley GF (2006) Ns-3 project goals. In: WNS2 '06 proceeding from the 2006 workshop on ns-2: the IP network simulator, Pisa

Heredia E (2008) An overview of the DLNA architecture, Redmond, Washington

Hohl L, Tellez R, Michel O, Ijspeert AJ (2006) Aibo and webots: simulation, wireless remote control and controller transfer. Rob Autom Syst 54(6):472–485

Holzner S (1993) The microsoft foundation class library programming. ISBN: 9781566861021

Honda (2010) Honda worldwide | ASIMO. http://world.honda.com/ASIMO/. Accessed 30 Noviembre 2010

Honda Motor Co (2010) Honda Worldwide | U3-X, Tokyo, Japan. http://world.honda.com/U3-X/. Accessed 1 Dec 2010

IFR International Federation of Robotics (2010) http://www.ifr.org/. Accessed 1 Dec 2010

Ingenium (2010) http://www.ingeniumsl.com/website/es/index.html. Accessed 2 Dec 2010

Intelligent Autonomous Systems Group (2010) http://ias.cs.tum.edu/. Accessed 23 Nov 2010

Intelligent Cooperative Systems Laboratory (2010) http://www.ics.t.u-tokyo.ac.jp/. Accessed 23 Nov 2010

Irobot (2010) Irobot cleaning robots. http://store.irobot.com/home/index.jsp. Accessed 1 Dec 2010

iRobot Corp (2000) Mobility robot integration software user's guide. http://www.irobot.com/

Jeronimo M (2004) It just works: UPnP in the digital home. http://www.artima.com/spontaneous/upnp_digihome.html. Accessed 12 Dec 2010

KA LawnBott (2010) Kyodo America – LawnBott. http://www.lawnbott.com/. Accessed 7 Dec 2010

Kaage U, Kahmann V, Jondral F (2001) An OMNeT++ TCP model. In: Proceedings of communication networks and distributed systems modeling and simulation conference, Prague, Czech Republic

Karakuri (2010) karakuri.info. http://www.karakuri.info/. Accessed 7 Dec 2010

Katholieke Universiteit Leuven Home Page (2010) http://www.kuleuven.be/english/. Accessed 20 Nov 2010

Kranz M, Rusu RB, Maldonado A, Beetz M, Schmidt A (2006) A player/stage system for contextaware intelligent environments. In: Proceedings of the System Support for Ubiquitous Computing Workshop (UbiSys 2006), at the 8th Annual Conference on Ubiquitous Computing (Ubicomp 2006), Orange County, California

LBNL's Network Research Group (2010) http://ee.lbl.gov/. Accessed 21 Nov 2010

Maestre JM, Camacho EF (2009) Smart home interoperability: the domoesi project approach. Int J Smart Home 3(3):31–44

Mellon Carnegie (2010) The robot hall of fame: unimate. http://www.robothalloffame.org/unimate.html. Accessed 3 Dec 2010

Members of the UPnP Forum (2008) UPnP device architecture 1.1. http://www.upnp.org/specs/arch/UPnP-arch-DeviceArchitecture-v1.1.pdf. Accessed 2 Dec 2010

Microsoft, MSDN Library (2010) UPnP concepts. http://msdn.microsoft.com/en-us/library/ms899570.aspx. Accessed 2 Dec 2010

Microsoft Robotics Developer Studio (2010) http://www.microsoft.com/robotics/. Accessed 1 Dec 2010

Microsoft Robotics Studio (2010) Documentation microsoft robotics studio. http://msdn.microsoft.com/en-us/robotics/cc136623. Accessed 24 Nov 2010

Mint Evolution Robotics (2010) Introducing Mint™. http://mintcleaner.com/. Accessed 1 Dec 2010

MIT media lab (2010) Huggable. http://robotic.media.mit.edu/projects/robots/huggable/overview/overview.html. Accessed 7 Dec 2010

MobileRobots (2010) MobileRobots research and academic customer support. http://robots.mobilerobots.com/wiki/Main_Page. Accessed 21 Nov 2010

Montemerlo M, Roy N, Thrun S (2003) Perspectives on standardization in mobile robot programming: the Carnegie Mellon navigation (CARMEN) toolkit. In: Proceedings of international conference on intelligent robots and systems, Las Vegas, USA, vol 1–4, pp 2436–2441. doi: 10.1109/IROS.2003.1249235

MoviRobotics (2010) http://www.movirobotics.com/. Accessed 2 Dec 2010

NASA New Millennium Program (2010) http://nmp.nasa.gov/. Accessed 7 Dec 2010

NATO, STANAG 4586 (Standard Interfaces of UAV Control System (UCS) for NATO UAV Interoperability), NATO Standardization Agency (NSA), Brussels (2004)

Nguyen HG (2005) Overview and highlights of robotics research and development at the space and naval warfare systems Center, San Diego. SPAWAR systems center. http://handle.dtic.mil/100.2/ADA433768. Accessed 1 Dec 2010

NI LabVIEW (2010) NI LabVIEW – Improving the productivity of engineers and scientists. http://www.ni.com/labview/. Accessed 12 Dicember 2010

NorthStar Evolution Robotics (2010) NorthStar system delivers robotic. http://www.evolution.com/products/northstar/. Accessed 7 Dec 2010

OMG (Object Management Group) (1998) CORBAservices: common object services specification. http://www.ing.iac.es/~docs/external/corba/CorbaServices.pdf. Accessed 2 Dec 2010

OMG's CORBA Website (2010) http://www.corba.org/. Accessed 20 Nov 2010

Open Dynamics Engine – home (2010) http://www.ode.org/. Accessed 23 Nov 2010

OpenGL (2010) OpenGL – The industry standard for high performance graphics. http://www.opengl.org/. Accessed 19 Nov 2010

OpenRDK Website (2010) http://openrdk.sourceforge.net/index.php. Accessed 21 Nov 2010

OPNET Home Page (2010) http://www.opnet.com/. Accessed 21 Nov 2010

Oracle (2010) When should I use RMI-IIOP? http://java.sun.com/j2se/1.5.0/docs/guide/rmi-iiop/rmiiiopUsing.html. Accessed 2 Dec 2010

Pal Robotics (2010) Pal robotics/humanoid robots. http://www.pal-robotics.com/. Accessed 1 Dec 2010

Parma University (2002) ARGO project home page. http://www.argo.ce.unipr.it/ARGO/english/index.html. Accessed 1 Dec 2010

Player Project (2010) http://playerstage.sourceforge.net/. Accessed 24 Nov 2010

Robobuilder (2010) Robobuilder: education and entertainment robotic DIY kit. http://www.robobuilder.net/eng/. Accessed 1 Dec 2010

RoboCup (2010) http://www.robocup.org/. Accessed 26 Nov 2010

Robomow (2010) Lawn mowers. http://www.robomow.com/. Accessed 7 Dec 2010

Rognlie R (1995) C++ robots introduction website with samples. http://www.pbm.com/~lindahl/pbem_articles/cpprobots_environment.html. Accessed 26 Nov 2010

RoSta (2010a) Architecture Patterns – RoSta. http://wiki.robot-standards.org/index.php/Comparison_and_Evaluation_of_Middleware_and_Architecture. Accessed 3 Dec 2010

RoSta (2010b) Middleware – RoSta. http://wiki.robot-standards.org/index.php/Middleware. Accessed 25 Nov 2010

SAE (2010) Society of Automotive Engineers. http://www.sae.org/. Accessed 2 Dec 2010

SAE, JAUS History and Domain Model (2006) Architecture framework committee. http://www.sae.org/. Accessed 2 Dec 2010

SAE, JAUS Mobility Service Set (2009) Information Modeling and Definition Committee. http://www.sae.org/. Accessed 3 Dec 2010

SAE, JAUS/SDP Transport Specification (2009) Network Environmental Committee. http://www.sae.org/. Accessed 5 Dec 2010

SAMAN (2001) Simulation augmented by measurement and analysis for networks (SAMAN). http://www.isi.edu/saman/index.html. Accessed 21 Nov 2010

Samsung (2010) SAMSUNG España http://www.samsung.com/. Accessed 7 Dec 2010

Schlenoff C, Albus J, Messina E, Barbera AJ, Madhavan R (2006) Using 4D/RCS toaAddress AI knowledge integration. AI Magazine, 27

Segway Inc (2010) Segway – The leader in personal, green transportation. http://www.segway.com/. Accessed 1 Dec 2010

Spykee World (2010) Spykee, the spy robot. http://spykeeworld.com/spykee/UK/index.html. Accessed 1 Dec 2010

Stepanov A, Lee M (1995) The standard template library

The ICSI Networking Group (2010) http://www.icir.org/. Accessed 22 Nov 2010

TORC. ByWire XGV (2010) – Hybrid escape drive-by-wire platform. http://www.torctech.com/. Accessed 3 Dec 2010

Universität Ulm Home Page (2010) http://www.uni-ulm.de/. Accessed 20 Nov 2010

Université de Sherbrooke Home Page (2010) http://www.usherbrooke.ca/. Accessed 21 Nov 2010

University of Oviedo. Infobotica Research Group (2010) DhCompliant stack architecture v0.4. http://156.35.46.38/data/files/Architecture/DHCompliantArchitecture0433.pdf. Accessed 3 Dec 2010

University of Seville (2010) Interoperability diagrams (technical report). http://nyquist.us.es/dhcompliant/Interoperability_diagrams.pdf. Accessed 3 Dec 2010

UPnP Forum (2001) Universal Plug and Play vendor's implementation guide. http://www.upnp.org/download/UPnP_Vendor_Implementation_Guide_Jan2001.htm. Accessed 12 Dec 2010

UPnP Forum (2010). http://www.upnp.org/. Accessed 2 Dec 2010

Urbi Home Page (2010) http://www.urbiforge.org/index.php/Main/HomePage. Accessed 26 Nov 2010

US Bureau of Labor Statistics (2010) Employment Projections Home Page. http://www.bls.gov/emp/. Accessed 1 Dec 2010

Utz H, Sablatnog S, Enderle S, Kraetzschmar G (2002) Miro – middleware for mobile robot applications. IEEE RSJ Trans Robot Autom 18:493–497

Varga A (2001) The OMNeT++ discrete event simulation system. In: Proceedings of the European simulation multiconference (ESM'2001), Prague, 2001

Varga A, Hornig R (2008) An overview of the OMNeT++ simulation environment. In: Proceedings of the 1st international conference on simulation tools and techniques for communications, networks and systems & workshops, Marseille, 2008

Vaughan RT, Gerkey BP, Howard A (2003) On device abstractions for portable, reusable robot code. In: Proceedings of the IEEE/RSJ international conference on intelligent robots and systems (IROS), Las Vegas, USA, 2003

Veizades J, Perkins C, Guttman E, Kaplan S (1997) Service location protocol. http://tools.ietf.org/search/rfc2165. Accessed 12 Dec 2010

VINT Project Website (1996) http://www.isi.edu/nsnam/vint/. Accessed 21 Nov 2010

W3C (2007) SOAP specifications http://www.w3.org/TR/soap/. Accessed 24 Nov 2010

W3C (2008) Guía Breve de Tecnologías XML. http://www.w3c.es/divulgacion/guiasbreves/tecnologiasxml. Accessed 12 Dec 2010

Wade RL (2006) Joint architecture for unmanned systems. Aviation and missile research, development and engineering Center (AMRDEC). http://www.dtic.mil/ndia/2006targets/Wade.pdf. Accessed 3 Dec 2010

Wehrle K, Reber J, Kahmann V (2001) A simulation suite for internet nodes with the ability to integrate arbitrary quality of service behavior. In: Proceedings of communication networks and distributed systems modeling and simulation conference, Phoenix, 2001

Wendel A, Bischoff K (2009) Robotics visions to 2020 and beyond. http://www.eurosfaire.prd.fr/7pc/doc/1286200019_g44_geoffpegman.pdf. Accessed 1 Dec 2010

Williams M (2002) PROMETHEUS-The European research programme for optimising the road transport system in Europe. In: Proceedings of the IEEE colloquium on driver information, p 1, London, UK

Willow Garage (2010) Overview | Willow Garage. http://www.willowgarage.com/pages/pr2/overview. Accessed 1 Dec 2010
WorldRobotic (2008) IFR/WorldRobotics. http://www.worldrobotics.org/index.php. Accessed 3 Dec 2010
WowWee (2010) WowWee™ astonishing imagination. http://www.wowwee.com/. Accessed 1 Dec 2010
XNA Developer Center (2010) http://msdn.microsoft.com/en-us/aa937791.aspx. Accessed 25 Nov 2010
Yujin Robot (2010) Iclebo. http://www.iclebo.com/product/iclebosmart.php. Accessed 29 Nov 2010
Zhang M, Zeigler BP, Hammonds P (2005) DEVS/RMI-An auto-adaptive and reconfigurable distributed simulation environment for engineering studies. ITEA J. http://citeseerx.ist.psu.edu/viewdoc/summary?doi=?doi=10.1.1.137.5872. Accessed 30 Nov 2010

Dissertations

Baity S (2005) Development of a next-generation experimentation robotic vehicle (NERV) that supports intelligent and autonomous systems research. Master of Science thesis, Mechanical Engineering, Virginia Tech, Virginia
Barrientos A (2002) Nuevas aplicaciones de la robótica. Robots de servicio. http://www.disa.bi.ehu.es/spanish/asignaturas/17219/Robots_Servicios-Barrientos.pdf
Bettstetter C, Renner C (2000) A comparison of service discovery protocols and implementation of the service location protocol. http://citeseerx.ist.psu.edu/viewdoc/summary?doi=10.1.1.37.3730
Certo G (2009) Nao Model and Simulation for Webots, Swiss Federal Institute of Technology, Lausanne
Cummings ML, Kirschbaum AR, Sulmistras A, Platts JT (2006) STANAG 4586 human supervisory control implications, Air and Weapon Systems Department, Dstl Farnborough & the Office of Naval Research
Farooq J, Bilal R (2006) Implementation and evaluation of IEEE 802.11 e wireless LAN in GloMoSim, Umea University, Umea
Faruque RR (2006) A JAUS toolkit for LabVIEW, and a series of implementation case studies with recommendations to the SAE AS-4 Standards Committee. Master of Science thesis, Mechanical Engineering, Virginia Tech, Virginia
Fielding RT (2000) Architectural styles and the design of network-based software architectures
Hidalgo Bláquez VM, Cañas JM (2008) Visual detection of vehicles speed in jdec platform
Levine J, Vickers L (2001) Robots controlled through web services: a technogenesis summer research. http://attila.stevens-tech.edu/webservices/robot.pdf
Lopez de Toro C, Ribas Xirgo L (2008) Anàlisi del, Microsoft Robotics Studio
Mojon S (2004) Using nonlinear oscillators to control the locomotion of a simulated biped robot. http://apl.epfl.ch/webdav/site/birg/shared/import/migration/diploma_report_mojon.pdf. Accessed 30 Nov 2010
Olleros GA (2007) Domotica: protocolo UPnP y Hogar Digital. Proyecto Fin de Carrera. Universidad de Sevilla, Sevilla. http://bibing.us.es/proyectos/abreproy/11557/fichero/Volumen+I%252F1_%CDndice.pdf. Accessed 27 Nov 2010
Santana JM (2005) Evaluación Del protocolo De Descubrimiento De Servicios Upnp EN redes Inalámbricas, Universidad De Malaga, Malaga
Satoshi K (2004) Cyberlink for java – Programming guide V 1.3.
Song H, Kim D, Lee K, Sung J (2005) UPnP-based sensor network management architecture, ICMU. http://www.ishilab.net/icmu2005/papers/117390-1-050228235605.pdf
Zhu F, Mutka M, Ni L (2002) Classification of service discovery in pervasive computing environments, Michigan State University, East Lansing

MIX
Papier aus verantwortungsvollen Quellen
Paper from responsible sources
FSC® C105338

If you have any concerns about our products,
you can contact us on
ProductSafety@springernature.com

In case Publisher is established outside the EU,
the EU authorized representative is:
Springer Nature Customer Service Center GmbH
Europaplatz 3, 69115 Heidelberg, Germany

Printed by Libri Plureos GmbH
in Hamburg, Germany